Making Math Meaningful™

9th Grade Workbook

By Andrew Starzynski and Jamie York

Jamie York Press

2019

Making Math Meaningful™

9th Grade Workbook

Text Copyright © 2012
by Andrew Starzynski and Jamie York

ISBN: 978-1-938210-10-5

All rights reserved. No part of this book may be reproduced in any form or by any means, without permission in writing from the publisher.

Cover Design Copyright © 2008 by Catherine Douglas.
Catherine Douglas was a student at Shining Mountain Waldorf School when she designed the cover to this book. It is an impressive example of how equiangular spirals emerge from nested octagons.

JAMIE YORK
PRESS

Meaningful Math Books for Waldorf,
Public, Private, and Home Schools
www.JamieYorkPress.com

To the Student

Welcome to high school! This year will be different in many ways from your previous educational experiences. You will have many different teachers, each one with a different style of teaching. Your first year of high school math studies will also take on a whole new level of seriousness. You will study algebra for most of the year. Algebra will form the foundation for all of your future mathematics studies; it is the "language of mathematics."

It does not matter how good or bad you think you are at mathematics – <u>EVERYONE</u> can do algebra! If you put forth good effort, stay positive, and ask the right questions, you will do well!

Here are some tips on how to use this workbook successfully:

- <u>Show your work</u> carefully and neatly on a separate sheet of paper. Make sure you write down the worksheet number and problem number, so you can easily find it later. All of this is important so that you can find your mistake if the answer comes out wrong. Algebra rewards those who are careful and organized with their work!

- Unless stated otherwise, all problems in this workbook should be solved <u>without the use of a calculator</u>.

- Above all, <u>homework is for learning</u>! Try your best on every problem. Struggling and overcoming frustration are part of the process of doing math. Even if you don't get a problem correct, you will learn by trying it, and then later seeing how it should be done. <u>Do not fall into the trap of doing the homework just to get it done</u>.

- <u>Don't give up!</u> If you get stuck on a problem, leave it and come back to it later. You will find that often it is easier to come to a solution when you return to a problem. If you can't solve it on the second try, ask your teacher or a classmate for help.

- <u>Learn from your mistakes</u>! When you get a problem wrong, make sure you follow up on it; find your mistake, and learn how to do the problem correctly.

Pre-Algebra

Problem Set #1

Vocabulary.
Here is some vocabulary you should be familiar with:
- Simplify
- Term
- Expression
- Product
- Sum
- Difference
- Base
- Exponent
- Operation
- Order of Operations
- Variable
- Equation

1) How many *terms* are in the following *expressions*:
   a) $4 + 7 + 10 - 11$ 4
   b) $5 - 3x + 2y$ 3
   c) $\frac{1}{x} + 4y = 8z$ 3
   d) $45x + 3y = 8z - 17x^3$ 4

2) What is the *product* of:
   a) 5 and 9   45
   b) 9 and 5   45
   c) ½ and 36   18
   d) 8 and 1   8
   e) 0 and 1,000,000   0

Simplify.
3) $15 - 3$ 12
4) $-3 + 15$ 12
5) $3 + (-15)$ -12
6) $-15 + 3$ -12
7) $15(3)$ 45
8) $(3)15$ 45
9) $(3)(-15)$ -45
10) $-3(15)$ -45
11) $-\frac{1}{4} + \frac{2}{7}$ $-\frac{7}{28} + \frac{8}{28}$ $\frac{1}{28}$
12) $\frac{1}{4} - \frac{2}{7}$ $\frac{7}{28} - \frac{8}{28}$ $-\frac{1}{28}$ $\frac{15}{28}$
13) $\frac{1}{4} - -\frac{2}{7}$ $\frac{1}{28}$
14) $-\frac{1}{4} - -\frac{2}{7}$ $\frac{1}{28}$
15) $\frac{1}{4} + -\frac{2}{7}$ $-\frac{1}{28}$
16) $\frac{1}{4} - +\frac{2}{7}$ $-\frac{1}{28}$
17) $\frac{3}{5} \cdot \frac{2}{3}$ $\frac{2}{5}$
18) $-\frac{3}{5} \cdot \frac{2}{3}$ $-\frac{2}{5}$
19) $\frac{3}{5}\left(-\frac{2}{3}\right)$ $-\frac{2}{5}$
20) $\frac{3}{5} \div \frac{2}{3}$ $2\frac{1}{2}$
21) $-\frac{3}{5} \div \frac{2}{3}$ $-2\frac{1}{2}$

Word Problems.

22) The product of 5 and some number is 40. Find the number.   8

23) The product of two numbers is 64. If those two numbers are equal, find the numbers. Is this the only answer?   8   I think so

— Pre-Algebra —
Problem Set #2

Simplify.
1) $-23 + 11$
2) $11 - 23$
3) $-11 + 23$
4) $23 - 11$
5) $54 - (17)$
6) $54 - 17$
7) $-17 + 54$
8) $17 - 54$
9) $-54 + 17$
10) $-54 - (-17)$
11) $-54 + (-17)$
12) 10^5
13) $10 \cdot 5$
14) $3 \cdot 4^2$
15) $1 + 3 \cdot 2$
16) $(1+3) \cdot 2$
17) $\frac{1}{4} - \frac{1}{3}$
18) $-\frac{1}{3} + \frac{1}{4}$
19) $\frac{1}{3} \cdot \frac{1}{4}$
20) $-\frac{1}{3} \cdot \frac{1}{4}$
21) $\left(\frac{1}{3}\right)\left(-\frac{1}{4}\right)$
22) $\frac{1}{3} \div \left(-\frac{1}{4}\right)$
23) $\frac{1}{3} \div 4$
24) $\frac{1}{4} \div 4$
25) $4 \div \frac{1}{4}$
26) $\frac{-1}{9} \cdot \frac{2}{-8} \cdot \frac{3}{7} \cdot \frac{4}{6} \cdot \frac{5}{5} \cdot \frac{-6}{4} \cdot \frac{7}{3} \cdot \frac{-8}{2} \cdot \frac{9}{1}$

Exponents.
27) Identify both the *base* and *exponent*:
   a) 12^5
   b) 3^x
   c) x^3
   d) $(½)^8$
28) Simplify the following:
   a) 3^2
   b) $3 \cdot 2$
   c) 9^4
   d) $9(4)$
   e) $(½)^3$
   f) $(½) \cdot 3$

Ordered Lists.
29) Write the following list in order from least to greatest:
   $-4, 4, -5, 5, -6, 6$

Word Problems.
30) The sum of five and a number is 15. Find the number.
31) The difference of a number and 12 is 13. Find the number. How many answers are there to this problem?
32) The product of 45 and a number is 135. Find the number.
33) Two with an exponent is 32. Find the exponent.
34) A base with an exponent of 2 is 25. Find the base. (There are two answers to this problem. Find them both.)

— Pre-Algebra —

Problem Set #3

Simplify.
1) $23 - 34$
2) $34 - 23$
3) $-8 + 14$
4) $14 - 8$
5) $-33 - 10$
6) $35 + (-10)$
7) $-10 - -35$
8) 3^3
9) $3 \cdot 3$
10) $(-2)^2$
11) -2^2
12) $\left(\frac{3}{4}\right)^3$
13) $\frac{3}{7} \cdot 7$
14) $7\left(\frac{3}{7}\right)$
15) $7 \div \frac{3}{7}$
16) $\frac{3}{7} \div 7$
17) $\frac{3}{7} \div \frac{3}{7}$
18) $\frac{3}{4} + \frac{4}{7} - \frac{5}{28}$
19) $(-3)(-2)$
20) $8(-3)$
21) $4 \div -2$
22) $\frac{4}{-2}$
23) $\frac{-4}{2}$
24) $9 \cdot 8 \cdot 7 \cdot 6 \cdot 5 \cdot 4 \cdot 3 \cdot 2 \cdot 1 \cdot 0$
25) $4^2 + 5 \cdot 8$
26) $9 + 18 \div 9$
27) $4 \cdot 10^2$
28) $(4 \cdot 10)^2$

29) $(3+4)^3$
30) $3^3 + 4^3$
31) $1 - 8(2+3^2)$
32) $120 \div 8 \div 2$
33) $8 + 3(2 - 2 \cdot 3^2) \div -8$
34) $\frac{1}{2} + \frac{3}{4} \cdot \frac{1}{5}$
35) $4\left(\frac{1}{3} + \frac{2}{9}\right) + 1$
36) $-1(3+4)$
37) $-(3+4)$
38) $(-1)(3)^2$
39) $(-1 \cdot 3)^2$
40) -3^2
41) $(-3)^2$

Word Problems.
42) The difference between four and twice a number is 2. Find the number.
43) The product of two equal numbers is 9. Find the numbers.
44) The sum of one-half and a number is $7/2$. Find the number.

Ordered Lists.
45) Write the following lists in order from least to greatest:

   a) $\frac{1}{2}, \frac{1}{3}, \frac{2}{3}, \frac{3}{5}, \frac{1}{9}, \frac{3}{2}, \frac{7}{4}, 1, 0$

   b) $3.1, 3.0, 3.006, 3.008, 3.02$

   c) $3, 3^2, \frac{1}{3}, \left(\frac{1}{3}\right)^2,$
   $2, 2^2, \frac{1}{2}, \left(\frac{1}{2}\right)^2$

— Pre-Algebra —

Problem Set #4

Order of Operations.
Simplify.
1) Use the *Order of Operations* to simplify the following:
   a) $2 - (4 + 3)^2$
   b) $(3 + 7)^2 \cdot 9 - 3$
   c) $3 \cdot 3^3 - 4(1 + 2)$
   d) $(3 + 2)^2 \div 5$
   e) $(3(4+2)^2) - 3^2 + 3 \cdot 8$
   f) $\frac{1}{2} - \frac{2}{3} \cdot \frac{4^2}{7}$

Fill in the blank.
2) a) If $x = 9$, then $x+7 = $ ___
   b) If $y = 3$, then $y-5 = $ ___
   c) If $a = -2$, then $5a = $ ___
   d) If $b = 12$, then $\frac{144}{b} = $ ___
   e) If $c = 4$, then $c^2 + 2c = $ ___
   f) If $d = -3$, then $7 - d^3 = $ ___

Formulas.
3) Use the formulas
$$C = \tfrac{5}{9} \cdot (F - 32)$$
$$F = \tfrac{9}{5} \cdot C + 32$$
Convert between Fahrenheit and Celsius
   a) Convert 90°F to °C
   b) Convert 50°C to °F
   c) Convert 0°F to °C
   d) Convert -40°C to °F

Solving Equations.
Simple *equations* can be easy to solve in our heads. For example: The solution to $x + 7 = 9$ is $x = 2$.

Solve for the variable:
4) $5x = 25$
5) $10 + y = 100$
6) $\frac{z}{2} = 4$
7) $7a = -63$
8) $5b + 3 = 33$
9) $c^2 = 4$
10) $z - 18 = -20$
11) $18x = 0$
12) $18x = 18$
13) $5x + 5 = 2x + 8$

Ordered Lists.
Write the following lists in order from least to greatest:
14) $\frac{5}{6}, 1, -1, \frac{-5}{6}, \frac{5}{-6}, \frac{-5}{-6}, \frac{6}{5}$,
$\left(\frac{5}{6}\right)^2, \left(\frac{-5}{6}\right)^2, \left(\frac{6}{5}\right)^2$
15) 5.01, 5.019, 5.021, 5.009, 5.0909

(This problem set is continued on the next page→)

— Pre-Algebra —

Word Problems.
16) The sum of a number and 5 is equal to 53. Find the number.

17) The product of eight and a number is 96. Find the number.

18) The difference of a number and 23 is 40. Find the number.

19) The sum of three consecutive (numbers next to each other) positive whole numbers is 30. Find all three numbers.

20) The sum of twice a number and 9 is 15. Find the number.

21) If 3 with an exponent is 27, find the exponent.

Problem Set #5

Formulas.
1) Use *Galileo's Law of Falling Bodies*:
$d = 4.90t^2$ (meters), or
$d = 16.1t^2$ (feet)
to calculate the distance an object falls after being dropped (neglecting air resistance)…
   a) For 1 second.
   b) For 5 seconds.
   c) For $2\frac{3}{5}$ seconds.
   d) For $7\frac{1}{4}$ seconds.

2) Convert between Fahrenheit and Celsius.
   a) 0°C to °F
   b) -40°F to °C
   c) $2\frac{7}{9}$°C to °F
   d) 16.2°F to °C

Fill in the blank.
3) a) If $x = -3$, then $5x + 2 =$ ___
   b) If $y = 4$, then $-5y - 2 =$ ___.
   c) If $z = 1.1$, then $z^2 =$ ___.
   d) If $a = \frac{1}{2}$, then $a + \frac{1}{3} =$ ___.

Simplify.
4) $10 \cdot 0.1$

5) $(-8)(7)(-2)$

6) $4 - 3 - (-4) + (-5) - -(-1)$

7) $3\left(\frac{1}{9}\right)$

8) $\frac{1}{3} \cdot \frac{1}{9}$

9) $\frac{3}{4} \cdot \frac{16}{9}$

10) $(1.1)^2$

11) $\left(\frac{11}{10}\right)^2$

12) $\frac{1}{5} + \frac{2}{3} - \frac{11}{15} \cdot \frac{1}{2}$

13) $\left(\frac{2}{3}\right)^3 - 4 \cdot 3^2$

14) $-3 - (-3) + (-3) - (+3)$

15) $(-3)(4) - (-1)(5)$

16) $\frac{-3}{4} - \frac{3}{-4}$

17) $\left(\frac{1}{3}\right)\left(\frac{9}{3}\right)\left(\frac{12}{4}\right)\left(\frac{7}{21}\right)$

Solving Equations.

18) Solve for the variable:
   a) $x + 3 = 17$
   b) $-7a = 84$
   c) $5y + 3 = 8$
   d) $\frac{b}{12} = 12$
   e) $\frac{12}{b} = 12$
   f) $\frac{1}{12}c = 12$
   g) $\sqrt{z} = 13$
   h) $3x + 2 = 4x - 1$

Combining Like Terms.

19) Simplify the following *expressions* by *combining like terms*:
   a) $5x + 3y - 8x$
   b) $7x^2 + 3x - 8y + 12x^2 - 4x$
   c) $y - x + 5 - xy + yx + 4xy + 18$
   d) $5a^2b + 6b^2a - 7ab + 15ab^2 - 8a^2b^2$
   e) $x - y + 42 - 2x + 3y - 3x - 2y - 2 + x^2$

Ordered Lists.

20) Write the following lists in order from least to greatest:
   a) $(-2)^2, -2^2, 2^2, (-2)^3, -2^3, 2^3, (-2)^4, -2^4, 2^4$
   b) $1, (-1)^2, (-1)^4, (-1)^6, (-1)^{1000}, -1^2, -1^4, (-1)^3$
   c) $-1, -2, -1.5, -1.75, -1.705, -½, -1.7005, 0$

Word Problems.

21) Three times a number is 81. Find the number.
22) The sum of twice a number and 40 is 80. Find the number.
23) One-half of the product of a number and two is 12. Find the number.

— Pre-Algebra —

Problem Set #6

Combine Like Terms.

1) What is the difference between an *expression* and an *equation*?
2) Simplify the following *expressions* by combining like terms:
   a) $5x + 3x - 4y + 8x$
   b) $1 + 4x^2 + 3x^2 - 7y^2 + 8y - 4$
   c) $5 - 3x^2 + 5x^3 - 76x^2 + 3y$
   d) $5x^2y + 6yx^2 - 8xy$
   e) $4xz + 3xy + 8xy^2z - 3yx^2z$
   f) $5x^3 - 3x^2$

Solve.

3) $2x + x = 6$
4) $7x - 5 + 12x + 7 = 40$
5) $4x + 15 - 12x = 7$
6) $½ x + ^3/_2 x + 3 = 9$
7) $4 = 2x + 3x - 13x$
8) $2x + 1 = 5x - 2$

Formulas.

9) Convert 82°F to °C
10) Convert -4°C to °F
11) How far has an object fallen after $4\frac{2}{5}$ seconds (neglecting wind resistance)? Give your answer in both meters and feet.

Fill in the blank.

12) a) If $b = 0.6$, then $\frac{3}{5} - b = $ ___.
    b) If $x = 4$ and $y = 1$, then $2x - 3y = $ ___
    c) If $c = -8$, then $c^2 = $ ___.
    d) If $d = -8$, then $-d^2 = $ ___.
    e) If $m = -8$, then $1 - m = $ ___.

Simplify.

13) $5(2(2^2)^2) - 4^2$
14) $3 + 2 \cdot 3^2 - 4(5 + 2^2)$
15) $6 - (9 - 11)^2$
16) $\frac{63}{80} \cdot \frac{64}{21}$
17) $8^2 - 3(5 - 10 \cdot 7^2)$
18) $(3 - 4)(5 + 6)^2$
19) $\frac{5}{6} + \frac{7}{8} - \frac{5}{9}$
20) $\frac{1}{13} \cdot \frac{7}{11}$
21) $-18 - 34 - 56 + 200$
22) $1^2 - 2^2 + 3^2 - 4^2 + 5^2$
23) $\frac{-3}{5} \cdot \frac{10}{6}$
24) $\frac{3}{-5} \cdot \frac{-10}{6}$

Word Problems.

25) Three times the sum of four and a number is 15. Find the number.
26) 16 is the product of negative eight and a number. Find the number.
27) The square of a number is 49. Find the number. Is this the only answer?
28) The difference of 5 and a number is 4. Find the number. Is this the only answer?

— Pre-Algebra —
Problem Set #7

Combine Like Terms.
1) Simplify by *combining like terms*:
   a) $5x + 3x + 4y$
   b) $-4x + 5x^2$
   c) $7a + 5b - 9a - 12c$
   d) $8x^3 + 3x^2 - 14x^3 - 2x + 3$
   e) $\frac{1}{2}x^2 - 2x^2 + 8y - a + 7a$
   f) $-14b + b - 8b$
   g) $8x + 3y$

Solve.
2) $3x = 9$
3) $8x + 2 = 4$
4) $8x + 2 = -4$
5) $\frac{1}{2}x = -12$
6) $\frac{x}{2} = -12$
7) $3x + 8 = 8x - 17$
8) $5 - 2x = 10x - 19$
9) $5x + 3 - 8x + 12 = 23x + 2$
10) $\frac{1}{2}x + 17 = 4x + 10$
11) $3x + 5 + 7x - 12 = 8x - 2 + 9x + 1$

Simplify.
12) $6 - 3(4^2 + 2)$
13) $\left(\frac{2^2}{3}\right)^2 - 8 \cdot 2^2$
14) $3(3(3(3(3(3(4 - 2^2))))))$
15) $4 - (-2)^2$
16) $4 - -2^2$

Evaluate.
17) Evaluate each, given that $x = 2$, $y = -1$ and $z = \frac{1}{2}$.
    a) $x + 2y$
    b) $4z + x + y^2$
    c) $z + z^2 + \frac{1}{4}$
    d) $4 - x - y$
    e) $4 - x - y^2$
    f) $4 - x + (-y)^2$
    g) $7xy + zx$
    h) $3(x + y)^2 + y^x$

Word Problems.
18) Twice the square of a number is 128. Find the number.
19) A number is the sum of three and twice one-third of itself.
20) Two with an exponent is 128. Find the exponent.
21) A base with an exponent of 2 is 1. Find the base.
22) Three times the sum of a number and 1 is 21. Find the number.

Algebra Basics

Problem Set #1

Section A
Signed Numbers
Simplify.
1) $-9 + 25$
2) $-9 - 25$
3) $23 - 31$
4) $-31 + 23$
5) $(5)(7)$
6) $(5)(-7)$
7) $-3 + 9$
8) $(-3)(+9)$
9) $3 - 9$
10) $(-3)(-9)$
11) $-3 - 9$
12) $(-15) \div (-5)$
13) $(15) \div (-5)$
14) $\frac{15}{-5}$
15) $3 - -8$
16) $7 - +11$
17) $-4 - -9$
18) $-7 - (-2 - 10)$

Expressions
Simplify.
19) $5X + 9X$
20) $2A + 8X - 8A$
21) $2 + 5X - 7$
22) $-7 - X - 12 - X - 3Y$
23) $3X - 21 + X$
24) $-5X + 1 - 5X - 1$
25) $X - 2Y - 7X + 16$

Equations
Solve each equation by getting X alone. Use the method of "moving terms" instead of showing how sides balance.

26) $2X + 6 = 7X - 4$
27) $8X + 3 = 4X - 11$
28) $4X - 3 + 10X = 7 + 2X - 58$

Section B
Solve.
29) $5X - 7X + 13 - 2X = -10 - 3 + X + 3$
30) $-9 - X + 4 + 12X - 6 - 3X - 12X = 1$
31) $X + 2X + 3X + 4 + 5X - 20X = 4$
32) $3(X + 5) = 8 - (X + 2)$

— Algebra Basics —

Problem Set #2

Section A

Signed Numbers
Simplify.
1) $(-4)(-7)$
2) $-4-7$
3) $(-5)(8)$
4) $\frac{-30}{-3}$
5) $\frac{30}{-3}$
6) $\frac{-30}{3}$
7) $15 + -9$
8) $-7 - -5$
9) $-2 - +9 - -7 - +4$

Order of Operations
Simplify.
10) $5 + 4 \cdot 2$
11) $(5 + 4) \cdot 2$
12) $7 - 5 \cdot 3$
13) $(7 - 5) \cdot 3$
14) $10 \cdot 5^2$
15) $(10 \cdot 5)^2$
16) $5 + 20 \div 4$
17) $18 \div 12 \div 2 - 2 \cdot 4$

Equations Solve.

18) a) $-3X = 24$ b) $-3X = -24$
19) a) $X + 4 = -18$ b) $X - 4 = 18$
20) a) $X \div 3 = 21$ b) $\frac{X}{3} = 21$ c) $\frac{1}{3}X = 21$
21) a) $-36X = -4$ b) $-4X = -36$
22) a) $\frac{3}{7}X = \frac{4}{5}$ b) $\frac{4}{5}X = \frac{3}{7}$
23) a) $-\frac{5}{7}X = \frac{5}{14}$ b) $X - \frac{5}{7} = \frac{5}{14}$

Section B

Solve.

24) $7X - 3 + 9X - 12 - 13X + 5 - X + 12 = 0$
25) $6 + 2(3X - 7) + 3X = 8 - 3(X - 4) - X + 5$
26) $\frac{1}{4}X - 3X + 3 = \frac{2}{5}X - 6$

— Algebra Basics —

Problem Set #3

Section A
Simplify.
1) $X + X$ (2x)
2) $X \cdot X$ x^2
3) $X \cdot X \cdot X \cdot X \cdot X$
4) $X \div X$
5) $5X - B + X - B - Y$
6) $-3X - 7 - X + 9$
7) $-8 - 2 + 6 - 7 + 4$
8) $-5 + -9X - +7 - -2X$
9) $(-4)^2$
10) $(-4)^3$
11) $(-4)^4$
12) $30 \div 8 \div 4$
13) $10 - 8 \cdot 10^3 \div 4 \cdot 2$
14) Which fraction isn't equal to the others?
   (a) $\frac{3}{-7}$ (b) $\frac{-3}{7}$ (c) $\frac{-3}{-7}$ (d) $-\frac{3}{7}$

Evaluate each expression given
$$X = -2;\ Y = -10;\ Z = -5$$
15) $X^2 + 2Y - 3Z$
16) $Y^2 - 5Z$
17) $-Y^2 - 5Z$
18) $4X - 2YZ + 3Z^2$

Solve.
19) a) $-X - 5 = -1$
    b) $-6X + 3 = -15X$
20) $36X + 7 = 12X - 5$
21) a) $\frac{X}{-5} = -30$
    b) $\frac{3}{5}X = -9$
22) $3(X + 2) + 5 = 1 - (X + 1)$

Section B
Solve.
23) a) $-4X = -\frac{2}{5}$
    b) $\frac{8}{9} = \frac{12}{X}$
24) a) $\frac{8X}{15} = -\frac{12}{5}$
    b) $\frac{-2}{X} = \frac{3}{X-5}$

25) $4X - 8 - 10 - 6X = -7 - 3X - 3 + 22$
26) $4X + 4 + 2(X - 3) = 10 - 6(3X + 4) + 5 - (4X - 7)$
27) $1\frac{1}{3}X - 3\frac{1}{4} = 5X + 4\frac{1}{2}$

Problem Set #4

Section A

Simplify.

1) $12 \div 3 + 6 \cdot 2$
2) $8 + 10 \div 6 + 2 \cdot 2^2$
3) $9 - 7(9 - 5 \cdot 2)^2 - 5$
4) $8(537 - 530) - 8$
5) $8(2X - 4) - 3$

Evaluate each expression given $X = -1$; $Y = 3$; $Z = 0$.

6) $Z^3 - 4XZ^2$
7) $7X^2Y^3 - 8X^3$
8) $X^9 + 23Z^5 - 6(Z - 10)$

Unusual solutions. Solve.

9) $8 - 2X = 7X - 6 - 9X$
10) $8 - 2X = 5 + 3X + 3$
11) $8 - 2X - 5 = -X + 3 - X$

Solve.

12) a) $-6X = -42$ b) $-28X = 4$
13) a) $-5 + X = -7\frac{1}{7}$ b) $-5X = -7\frac{1}{7}$
14) a) $8X + 3 = 12X - 13$ b) $5\frac{1}{2}X + 3 = 7 - \frac{7}{2}X$
15) a) $5(X - 3) = -3(X - 1)$ b) $4(X + 3) - 1 = 1 - (2X + 3)$

Section B

Solve.

16) a) $\frac{-2}{5} = \frac{-3}{5X}$ b) $\frac{3X}{7} = \frac{-3}{8}$ c) $-1\frac{3}{5}X = \frac{4}{7}$

17) a) $\frac{-3}{X-5} = \frac{2}{3X+2}$ b) $2X - \frac{3}{4} = \frac{3}{4}X + 3$

18) $7 - 3(2X - 7) + 5X = 5X - 20 + 4(X - 8) + 8 - 2X$

19) $-\frac{1}{3}X + \frac{2}{3} - \frac{1}{4}X = \frac{7}{10} + \frac{5}{6}X - \frac{2}{5}$

— Algebra Basics —

Problem Set #5

Section A
Simplify.
1) $4 + 3 \cdot 9$
2) $6 - 5 \cdot 3 + 20$
3) $7 \cdot 3 + 12 \div (9 - 10)$
4) $30 - 10 \cdot 3^2$

Evaluate each expression given that x = 3; y = –2.
5) $5y - 6x + 3$
6) $y^2 - xy + 4 - \left(\frac{x}{y}\right)^3$

Unusual solutions. Solve.
7) $12 + 3(2X - 4) = 6X$
8) $6 = 9 - (4X + 3) - 2X$
9) $\frac{12}{4X+3} = \frac{3}{X-3}$

Solve.
10) a) $6X = -\frac{4}{5}$ b) $6 - X = -\frac{4}{5}$
11) a) $-2\frac{2}{3}X = -\frac{4}{7}$ b) $X - 2\frac{2}{3} = -\frac{4}{7}$
12) a) $-X = -13$ b) $5 - 2(X + 4) = 0$
13) a) $\frac{-5}{3X+1} = \frac{2}{2X-3}$ b) $-8X + 3 - 5X = 7 + 2(X - 7)$
14) $5 + 3(X - 2) - 4 = 6 - 5(2X - 3)$

Section B
Solve.
15) $\frac{7}{9} - \frac{4}{9}(6X - \frac{3}{4}) = \frac{3}{5}X - \frac{1}{3}$
16) $4X - 7 - 4(X+5) + 5 - 3(4X-2) - 6 = 3X - 30 + 9(X-2) - 8$

Problem Set #6

Section A
Solve for x in terms of Y.
1) $X + 5Y = 4$
2) $4X = 3Y$
3) $3Y = 3 + X$
4) $3X - 12Y = 9$
5) $3X + 7Y = 5$

Solve.
6) $5X + 3 = 12X - 67$
7) $14 - (3X + 2) = 3X$
8) $\frac{-6}{2X+5} = \frac{-4}{4X-3}$
9) $\frac{1}{5}X - 3 = \frac{2}{5}(X + 1)$
10) $3(X - 2) + 1 = 4X - 2$

— Algebra Basics —

Section B

Solve.

11) $8 + 2(3X - 5) - X - 4(X + 7) = 5 - (2X - 7) + 8(3 - 2X)$

12) $\frac{2}{5}(2X - \frac{1}{2}) = \frac{2}{3}X + \frac{1}{3}$

13) $\frac{2}{5} + \frac{1}{2}(\frac{4}{5}X - 1) = -\frac{1}{5}(\frac{5}{6}X - 2\frac{1}{2}) - 1\frac{1}{2}$

Problem Set #7

Section A

Solve for x in terms of Y.

1) $2X + Y = 8$
2) $Y = 4X - 12$
3) $Y = \frac{2}{3}X + 5$
4) $3Y + 7X = -3$
5) $\frac{3}{4}X - Y = -3$
6) Solve for C
   $F = \frac{9}{5}C + 32$

Solve.

7) $5X + 1 = 7X - 8$
8) $3X - 4(X + 2) = 5 - 5(2X - 1)$
9) $2\frac{1}{3}X + 3 = 5X - 1\frac{5}{6}$
10) $\frac{1}{5x+2} = \frac{2}{x+1}$

Section B

Solve.

11) $5X + 5 - 2(X - 3) - 5 - 6(-X - 2) - 1 = 3X - 30 - (4X - 7) - 8$

12) $\frac{5X + 6}{2} = \frac{2X + 9}{3}$

13) $\frac{4}{7} - \frac{3}{7}(\frac{2}{3}X - 3) = \frac{3}{8} - 4(\frac{4}{7}X - \frac{3}{5}) + 2X$

14) $\frac{2}{5}X - 8\frac{1}{8} - \frac{3}{4}(\frac{14}{15}X - 5\frac{5}{9}) = 2\frac{7}{10}X - 2\frac{5}{6} + \frac{3}{4}(X + 5\frac{1}{6})$

— Algebra Basics —

Problem Set #8
Section A

1) X is a grade on a hundred-point scale and Y is a grade on a four-point scale. For example, a "B" is given as X = 85 and Y = 3.0. Also, we know that if X = 90 then Y = 3.5.
   a) What is the formula for X in terms of Y?
   b) What is the formula for Y in terms of X?

Solve for x in terms of Y.

2) $5X - 3Y = 15$

3) $Y = \frac{3X - 4}{5}$

4) Solve for F
   $C = \frac{5}{9}(F - 32)$

Solve.

5) $5(2X - 4) = 18 - 2(3X + 1) - 4$

6) $\frac{-4}{x+2} = \frac{5}{2x+1}$

7) $-\frac{4}{5}X - 5 = \frac{1}{2} - 3X$

8) $5\frac{1}{2}X - \frac{2}{3} = \frac{3}{4}X$

9) $-3X - 5 - 8X - 3 = 7X - 4 - 10X$

Section B
Solve.

10) $\frac{5}{8} - \frac{1}{8}(\frac{1}{3}X + \frac{8}{9}) - X = \frac{5}{24}X + 2\frac{1}{2}(\frac{3}{4}X + \frac{1}{4}) - \frac{5}{9}$

11) $\frac{2}{3} - \frac{1}{6}(2X - 3) + \frac{X}{4} = \frac{3}{8}X + \frac{1}{5}(\frac{5}{6}X - 2) + 3\frac{1}{2}$

Problem Set #9
Section A
Evaluate each expression given that $a = \frac{2}{3}$; $b = -\frac{3}{4}$.

1) $3b - 4a + 3$

2) $6a^2$

3) $\frac{4}{a^2}$

4) $\frac{5}{4} - b \cdot a$

5) $-b^2 + \frac{1}{a}$

6) $\frac{1}{b^2} - (6ab)^3 - \frac{1}{2} - \frac{a}{b}$

Solve for x in terms of Y.

7) $5Y - 6X = 4$

8) $Y = \frac{1}{3}X + 1\frac{2}{3}$

9) $\frac{X}{7} - \frac{5}{7}Y = \frac{3}{7}$

10) $\frac{5}{6}Y + \frac{3}{4}X = -\frac{2}{3}$

Solve.

11) $5(3X - 2) = 4(5 - 17X) + 10$

12) $\frac{1}{7}X + 3 = \frac{2}{21}(21X - 1)$

13) $\frac{4}{5x+1} = \frac{-3}{2x-1}$

14) $6X - (4X + 5) - 7(-X + 3) = 15X - 36 - 2(3X - 5)$

— Algebra Basics —

Section B.

15) a) $\frac{-1/2}{2/3 X+5} = \frac{-3/4}{1/4 X-3}$ b) $\frac{3}{8}\left(\frac{5}{6}X - \frac{2}{3}\right) = \frac{1}{3}\left(3X - \frac{3}{5}\right) - \frac{1}{20}$

16) $5 - \frac{3}{4}X + \frac{2}{7}\left(3X - \frac{7}{8}\right) = 5\frac{1}{2} - 2\frac{1}{2}\left(\frac{2}{5}X - 1\frac{1}{3}\right) + \frac{6}{7}X$

17) $\frac{3}{4}X + 9 - 2\left(\frac{3}{10}X - \frac{1}{6}\right) = \frac{1}{5}\left(\frac{2}{3} - X\right) - \frac{3}{4}\left(\frac{4}{5}X - 1\right)$

Problem Set #10

Section A
Evaluate each expression given that $a = 3/10$; $b = -1/3$.

1) $\frac{8}{10} - a \cdot b^2$

2) $10a^2$

3) $-\frac{1}{a^3}$

4) $20a^2 b^3 - ab + \frac{1}{b}$

5) $2(20a - 3b)^2 - \frac{5a}{b-1}$

Solve for C in terms of D.

6) $3D - 4C = 24$

7) $D = \frac{3}{4}C - 4\frac{1}{2}$

8) $\frac{C}{4} - \frac{5}{6}D = \frac{7}{12}$

9) $\frac{1}{5}C - D = \frac{4}{5}$

10) $2\frac{2}{3}D + 3\frac{1}{2}C = -\frac{5}{6}$

Solve.

11) $3(X - 2) + 1 = 4 - 3(2X - 1)$

12) $4\frac{1}{2}X - 3(X + 1) = \frac{1}{5}(2X + 1) + 3$

13) $\frac{-2}{3x-1} = \frac{-7}{2-x}$

14) $2X - 3(X - 4) + 3(2X + 5) = 5X + 8 - 3(3X + 1)$

Section B

Solve.

15) a) $\frac{4/11}{1/4 X + 2} = \frac{-2/3}{3X - 3\frac{2}{3}}$ b) $\frac{3}{5}\left(2\frac{1}{2}X - \frac{2}{3}\right) = \frac{1}{5}\left(10X - \frac{5}{6}\right) - \frac{5}{12}$

16) $\frac{4}{5} - \frac{1}{5}\left(\frac{5}{6}X - \frac{1}{2}\right) + 4X = \frac{2}{3}\left(8\frac{1}{4}X - 4\right) - 3\left(4\frac{1}{3}X - \frac{1}{4}\right) + 11\frac{1}{3}X$

17) $\frac{7}{9}X - 2\left(\frac{1}{3}X - \frac{5}{24}\right) = \frac{4}{5}\left(3\frac{1}{3} - \frac{5}{9}X\right) - \frac{3}{4}\left(\frac{4}{9}X - 1\frac{1}{2}\right)$

Exponents & Polynomials

Problem Set #1

Section A
Simplify.
1) $4x^3 + 7x^3$
2) $(4x^3)(7x^3)$
3) $7x + 8x$
4) $(7x)(8x)$
5) $\frac{2}{3}x^2 - \frac{3}{4}x^2$
6) $(\frac{2}{3}x^2)(-\frac{3}{4}x^2)$
7) $a^5 + a^5$
8) $(a^5)(a^5)$
9) $6w^5 - 20w^5$
10) $(6w^5)(20w^5)$
11) $3c^3 - 3c^3$
12) $(3c^3)(3c^3)$
13) $-6r^4 - 3r^4$
14) $(-6r^4)(-3r^4)$
15) $5x^3 + 2x^5$
16) $(5x^3)(2x^5)$
17) $-\frac{2}{7}y^3 - \frac{1}{2}y^3$
18) $(-\frac{2}{7}y^3)(-\frac{1}{2}y^3)$
19) $5x^3 + 6x^3 + 11x^3$
20) $5x^3 + 6x^2 + 11x^3$
21) $3a^5 - 2a^5 - 5a^5$
22) $3a^3 - 2a^4 - 5a^5$
23) $3x^5 - 2a^5 - 5y^5$
24) $(3a^5)(2a^5)(5a^5)$

Solve.
25) $5x + 2 = 3x - 7$
26) $4 - 7x = 7x + 4$
27) $8(9x - 2) + 1 = 7 - 2(x - 1)$
28) $3(2x + 3 - 4x) = -(4 - x) - 9$
29) $\frac{1}{3}x - 4 = 2 + \frac{2}{5}x$
30) $\frac{1}{2x+1} = 7$
31) $5 + 4(x - 8) = 2 - 2(2x - 1)$

Section B
Solve.
32) $5 - 2(3x - 8) + 4x - (x + 7) - 10x = 6(2x - 3) - 3(-7x - 8) - 46x + 12$

33) $\frac{1}{6} - \frac{2}{3}(9x + \frac{3}{5}) + \frac{3}{4}x = \frac{4}{15} - \frac{5}{6}(\frac{3}{10}x - 1\frac{14}{25})$

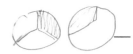

— Exponents & Polynomials —

Problem Set #2

Group Work
Vertical Multiplication

Example #1:
```
      52
     x38
     416
    1560
    1976
```

Example #2:
```
    5x + 2
    3x + 8
   40x + 16
  15x² + 6x
  15x² + 46x + 16
```

Multiply like the example.
1) a) 43 × 23
   b) $(4x + 3)(2x + 3)$
2) a) 32 × 21
   b) $(3x + 2)(2x + 1)$
3) a) 52 × 38
   b) $(5x + 2)(3x + 8)$
4) With #2 (above) both problems produce the same digits, but this is not the case for #1 and #3. Why is this so?

Section A
Simplify.
5) $x^4 + x^4$
6) $(x^4)(x^4)$
7) $x + x$
8) $x \cdot x$
9) $\frac{1}{3}x^5 - x^5$
10) $(\frac{1}{3}x^5)(-x^5)$
11) $-7w^5 - 3w^3$
12) $(-7w^5)(-3w^3)$
13) $4y^5 - y^5$
14) $\frac{6}{11}y^5 - \frac{1}{3}y^5$
15) $(\frac{6}{11}y^5)(-\frac{1}{3}y^5)$
16) $(4w^3)(2w^2)(10w^5)$
17) $5x^5 - 3x^4 + 6x^3 - 8x^5 - x^3$
18) $5x^2y^3 + 2x^2y^3$
19) $(5x^2y^3)(2x^2y^3)$
20) $4x^3y^2 + 3x^3y^5$
21) $(4x^3y^2)(3x^3y^5)$
22) $5x^3 - 4x^3y^5 + x^3 - 3x^3y^5$
23) $(5x^3)(-4x^3y^5)(x^3)(-3x^3y^5)$
24) $(5x^4y^3)^3$

Solve.
25) $3x + 2 - 4x = 8x - 1 + 13x - 5$
26) $5(x + 3) = 10x + 1 - 5x$
27) $1 - 4(3x + 6) = -14x + 5(3 - 5x)$
28) $\frac{2}{3x+1} = 5$
29) $\frac{1}{2}x - 3 = \frac{3}{7}x + 1$

Section B
Solve.
30) $7 - 5(2x - 3) + x - (3x - 2) - 8x = 3(x+3) - 2(-4x - 5) - 4x + 7$

31) $-7 - 3\frac{3}{5}(3\frac{1}{3}x - \frac{5}{6}) + 3\frac{2}{3}x = \frac{5}{8} - \frac{3}{8}(\frac{4}{5}x + \frac{2}{3}) + \frac{2}{15}x$

— Exponents & Polynomials —

Problem Set #3

Section A
Simplify.
1) $7x^2 - x^2$
2) $-7y^5 - 2y^5$
3) $x^3 + x^3$
4) $(x^3)(x^3)$
5) $5x^4y^3 - y^3$
6) $5x^4y^3 - 2x^4$
7) $5x^4y^3 - 2x^4y^4$
8) $(5x^4y^3)(-2x^4y^4)$
9) $5x^4y^3 - 2x^4y^3$
10) $3x^3 + x^2 + 7x^3$
11) $(3x^3)(x^2)(7x^3)$
12) $(3x^3y^5)^2$
13) $(10x^3y^2)^5$
14) $(3xy^4)^4$
15) $5(3x + 5)$
16) $5x^2(3x + 5)$
17) $5x^3(3x^4 + 5x^3)$
18) $4x^3(5x^2 - 6)$

Multiply out horizontally
19) $(5x + 1)(10x + 3)$
20) $(4x + 5)(4x + 5)$

21) $(3x + 4)(2x + 5)$
22) $(3x - 4)(2x + 5)$
23) $(3x + 4)(2x - 5)$
24) $(3x - 4)(2x - 5)$

Solve.
25) $5x - 34(x-1) = 5 + 3(x-7)$
26) $(x + 4)5 - 7(x - 1) = 2$
27) $\frac{1}{3}(x - 1) + 8 = \frac{2}{9}x$
28) $\frac{2}{x+1} = \frac{-3}{2x-1}$
29) $3\frac{1}{2}x + 2 = 5\frac{1}{4}x - 3$

Section B
Simplify.
30) $(5x^2)(6x^4y)(w^3xy^3)(5w^5)$
31) $\frac{5}{8}w^5 + \frac{1}{12}w^5$
32) $x^2 - 4x^3y + 2x^3 + 9x^3y$

Multiply.
33) $(6x + 5)(3x + 4)$
34) $(6x - 5)(3x - 4)$
35) $(6x + 5)(3x - 4)$
36) $(6x - 5)(3x + 4)$
37) $(7x^5 - 6)(2x^5 - 9)$

Solve.
38) $7 - 2(x+3) + 4x - 4(3x-2) - x = 5(2x-1) - 3(10x+3) + 3x + 8$

39) $3 - \frac{5}{6}\left(\frac{18}{35}x + 4\right) = 2\frac{2}{3} - \frac{1}{3}\left(\frac{4}{7}x + 11\frac{3}{4}\right) - x$

— Exponents & Polynomials —

Problem Set #4

Section A
Simplify.
1) $6x^3 + 7x^3$
2) $(6x^3)(7x^3)$
3) $(5x^4)(3x^3)$
4) $5x^4 + 3x^3$
5) $3x^3 - 2x^2 + x^3 - 7x^2$
6) $(3x^3)(-2x^2)(x^3)(-7x^2)$
7) $6(5x^2 + 3x - 8)$
8) $6x^3(5x^2 + 3x - 8)$
9) $6x^3(7x^3 + 5x^2 + 3x - 8)$
10) $2x^5(4x^6 - 3x^3 + 7)$
11) $5x^3y^2(3x^2 - 5y^5)$
12) $5x^4y^2 - y^2 + x^4y^2$
13) $(6x^3y^5)(2w^3y^8)$
14) $(5x^4)^2$
15) $(3x^3y^4z^3)^4$
16) $3x^3(2x^4)^2$
17) $3x^2y^3(10x^4y^3)^3$

Multiply.
18) $(3x + 2)(4x + 1)$
19) $(4x + 7)(3x + 5)$
20) $(x + 8)(x + 6)$
21) $(5x - 3)(3x - 2)$
22) $(5x + 3)(3x - 2)$
23) $(x + 8)(x - 3)$
24) $(x + 5)(x + 11)$
25) $(x - 5)(x + 11)$
26) $(x + 5)(x - 11)$
27) $(x - 5)(x - 11)$
28) $(4x + 5)^2$

29) The previous problem is exactly the same as which problem on the previous problem set? (Did you get the same answer?)

30) $(2x + 3)^2$
31) $(x + 7)^2$
32) $(5x - 6)^2$

Solve.
33) $\frac{2}{3}x + 8 = \frac{3}{4}x - 7$
34) $4 - (x - 1) = -12 - x$
35) $\frac{5}{2x+1} = \frac{-3}{5-x}$

Section B
Estimating Powers of Two
36) Given that $2^{10} = 1024$, we can estimate large powers of two by using $2^{10} \approx 1000$. For example, $2^{23} = 2^{10} \cdot 2^{10} \cdot 2^3 \approx 1000 \cdot 1000 \cdot 8$. Therefore we can say that $2^{23} \approx 8{,}000{,}000$.
Approximate each one:
  a) 2^{21}
  b) 2^{33}
  c) 2^{52}

Solve.
37) $8x + 3(-15 + 4x) = 5(x + 2x)$
38) $7 - 3(x + 1) = 6(8 - x) + 5$
39) $(x + 3)(x - 1) = (x + 7)(x - 4)$
40) $\frac{x+1}{x+2} = \frac{x+3}{x+4}$

41) $3\frac{1}{2}x + 4(x + 2\frac{1}{3}) - 3\frac{1}{4} = 5x + 17\frac{1}{2} - 4(5\frac{1}{3}x + 1)$

— Exponents & Polynomials —

Problem Set #5

Section A
Simplify.
1) $7x^4 - x^4$
2) $(7x^4)(-x^4)$
3) $3x^5 + 8x^3$
4) $(3x^5)(8x^3)$
5) $5x^2y^3 + 2x^2y^3$
6) $5x^2y^3 + 2x^2$
7) $3y^3 + 4x^3 - 7y^3 - 3x^2$
8) $8(3x^2 - 5x + 7)$
9) $4x^5(5x^3 - 4x^2 + x - 5)$

Multiply.
Try doing it in your head!
10) $(4x - 7)(3x - 2)$
11) $(10x + 3)(5x - 4)$
12) $(x + 3)(x + 6)$
13) $(x - 4)(x + 3)$
14) $(x - 5)(x - 8)$
15) $(x + 6)(x + 8)$
16) $(x + 6)(x - 8)$
17) $(x - 6)(x + 8)$
18) $(x - 6)(x - 8)$
19) $(x^3 + 6)(x^3 + 8)$
20) $(x + 6y)(x + 8y)$
21) $(5x - 2y)(x - 3y)$
22) $(x + 6)^2$
23) $(x - 4)^2$
24) $(x - 4y)^2$
25) $(x^5 - 4)^2$

Solve.
40) $(2x + 5)(3x - 1) = (6x + 1)(x + 7)$
41) $3 - \frac{1}{4}(\frac{1}{3}x + 18) - 2\frac{1}{4}x = -x + \frac{1}{4}(3x - 4\frac{2}{3}) - \frac{11}{18}$

26) **Evaluate.**
given $x = 4$; $y = -3$…
  a) $x^2 + xy - y + y^2$
  b) $2(x + y)^x + x(y + 1)^2$

Solve.
27) $\frac{5}{x} = \frac{9}{2x}$
28) $\frac{4+x}{3} = 2x - 24$
29) $\frac{5}{x} = \frac{x}{5}$
30) $(x - 3)(x - 2) = (x + 7)(x - 1)$

Section B
31) Powers of Two.
  a) $2^{34} \approx$ ___ b) $2^{42} \approx$ ___

32) Given $x = \frac{2}{3}$; $y = -\frac{1}{2}$
evaluate $xy^2 - \frac{1}{x} - \frac{x^2}{2y^3}$

Simplify.
33) $5n^3x^2(x^3 - 4n^3x^4 + 3n^2x^3)$
34) $k^3m^2p^3 - 6k^3m^2p^3$
35) $(5m^2q^3)(3m^6x^4)(2q^6y)$

Multiply.
36) $(5x^3 - 2y^2)(x^3 - 3y^2)$
37) $(4x^5 + 3y^4)^2$
38) $(7x^5y^2 - 10xy^4)^2$
39) $(x + 16)(x - 16)$

— Exponents & Polynomials —

Problem Set #6

Section A

Multiply.
Try doing it in your head!
1) $(7x - 3)(2x - 3)$
2) $(7x - 3)(2x + 3)$
3) $(3x - 7)(2x + 6)$
4) $(x + 3)(x + 2)$
5) $(x - 3)(x - 2)$
6) $(x + 3)(x - 2)$
7) $(x - 3)(x + 2)$
8) $5(x - 3)(x + 2)$
9) $5x^3(x - 3)(x + 2)$
10) $(x^2 - 6)(x^2 - 2)$
11) $(x + 6y)(x - 8y)$
12) $(x + 10)^2$
13) $(x - 1)^2$
14) $(2x + 3y)^2$
15) $(x^5 - 4)^2$

Multiplying a trinomial by a binomial.
Example:
$(2x + 3)(4x^2 + 7x + 2)$
Solution:
$$\begin{array}{r} 4x^2 + 7x + 2 \\ 2x + 3 \\ \hline 12x^2 + 21x + 6 \\ 8x^3 + 14x^2 + 4x \\ \hline 8x^3 + 26x^2 + 25x + 6 \end{array}$$

16) $(x - 6)(x^2 - 7x + 5)$

Simplify.
17) $5x^4 + x^4$
18) $(5x^4)(w^2x^4)$
19) $3x^5 - 8x^5$
20) $(5x^2y^3)(2x^2)$
21) $(3x^5)(-8x^5)$
22) $7a^3b^5c^2 + 6a^3b^5c^2$
23) $7a^3b^5c^2 + 6a^3b^5c^4$
24) $4(5x^3 - 4x^2 + x - 5)$
25) $(7a^3b^5c^2)(6a^3b^5c^4)$
26) $5x^4 - 2x^3 - 2x^4 + 6x^3$

Solve.
27) $(x + 5)(x + 3) = (x - 2)^2$
28) $\frac{4}{x} = \frac{x}{9}$

Section B

Simplify.
29) $6(5w^5 - 4w^3 + 3w - 5)$
30) $4z^5(z^3 + 5z^2 - 7z - 5)$
31) $x^6 - 5x^4 - 2x^4(x^2 - 1)$
32) $6x^2y^5(3x^4 - 5x^2y^2 + 6y^4)$
33) $(4x^3y^5)(3y^2z)(10x^3z^7)$

Multiply.
34) $(3x + 5)(x - 3)$
35) $(x^5 + 2y^3)(x^5 + 7y^3)$
36) $5x^2(x^5 - 4)^2$
37) $(x^2 - 4y^3)^2$
38) $(x + 4)(x + 3)(x+5)$
39) $(x + 5)^3$

Solve.
40) $(2x - 1)^2 = (4x + 3)(x + 4)$
41) $5x + (x+2)(9x-3) = (3x-1)^2$

— Exponents & Polynomials —

Problem Set #7

Section A

1) Leave answers with positive exponents, and with the square roots assume that x is positive.

a) $\sqrt{x^6}$ b) $\sqrt{16x^{16}}$

c) $\dfrac{x^7}{x^2}$ d) $\dfrac{x^2}{x^7}$

e) $\dfrac{x^{-4}}{x^{-6}}$ f) $\dfrac{3x^8 y^5}{6x^2 y^6}$

Simplify.

2) $3x^4 + 10x^4$
3) $(3x^4)(10x^4)$
4) $6a^4 d^2 - a^4 d^2$
5) $(6a^4 d^2)(-3a^4 d^2)$
6) $5x^2(3x^4 - 7x^2 + 3)$
7) $4x^3 y^5(3x^2 y^3 + 5x^2 - 7y)$
8) $x^5 - 3x^3 + 2x^2(x^3 - 4)$
9) $(½x^3 y)(5x^2 y^5)(8x)$
10) $(2x^5 y^2)^3$
11) $5x^2(3x^4 y)^2$
12) $(-3x^3 y^4)^3$
13) $(4x^3 y^2)^2 (-2xy^3)^4$
14) $(4x - 3)(2x + 5)$
15) $(x - 3)(x + 3)$
16) $(4x + 3y)(2x + 5y)$
17) $(4x - 3y)(2x + 5y)$
18) $(3x^2 + 1)(x^2 + 7)$
19) $(3x^2 + 1)(x + 7)$
20) $3(x - 5)(x + 6)$
21) $3x^4(x - 5)(x + 6)$
22) $(x - 7)^2$
23) $(x + 7)(x - 7)$
24) $(x - 4)^3$
25) given $x = 7$; $y = -½$
    Evaluate $y^2 - y(x^2 - 8y)$

Solve.

26) $5(x - 3) = 5 - (x - 3)$
27) $½x + 3 = ¾x - 2$
28) $\frac{1}{2}(x+2) = \frac{2}{3}(3x + 9)$
29) $\dfrac{4+x}{3} = \dfrac{3-x}{4}$
30) $(x + 2)^2 = (x - 7)^2$

Section B

Simplify.

31) $(½x + 6)^2$
32) $(4x^3 + 5y^2)(4x^3 - 5y^2)$
33) $(x - 4)(x + 1)(x - 3)$
34) $(x^2 - 3x + 2)(x^2 + 4x - 2)$
35) $\dfrac{6x^3 y^8}{15x^6 y^2}$
36) $\dfrac{(-2x^3 y^2)^3}{(-4xy^3)^2}$

Solve.

37) $(3x + 4)(4x + 3) = (6x - 2)(2x - 6)$
38) $(x + 1)^3 = x(x^2 + 3x - 4)$
39) $\dfrac{7}{12} - \dfrac{5}{12}\left(3\dfrac{9}{10}x - \dfrac{4}{5}\right) + 2x = \dfrac{3}{4}x - \dfrac{1}{12} - \left(\dfrac{5}{6}x - \dfrac{3}{16}\right)$

— Exponents & Polynomials —

Problem Set #8 (for groups!)

1) Fill in all of the tables on the next page starting with N = 1 and going down to N = 10.

2) Use the tables to answer the following questions:
   a) What is 3^7? 2187 b) What is 5^6? 15625
   c) What is 2^{10}? 1624 d) What is 10^5? 100000

3) On the three's table, every time you move down one step, the answer gets multiplied by 3. Answer the following:
   a) What happens when you move *down* one step on the five's table? It gets muliplied by 5
   b) What happens when you move *up* one step on the five's table? It gets dovided by 5
   c) Given that the five's table says that $5^1 = 5$, what is the answer when you move one step up to 5^0? 1 And another step up to 5^{-1}? .20 = $\frac{1}{5}$

4) Fill in each of the tables starting with N = 0 and going up to N = −5. Leave your answers as fractions. (You shouldn't have to do any calculations.)

5) Given what you now know, complete each of the following statements:
   a) Anything to the zero power equals... 1
   b) Anything to a negative exponent is the same as... a fraction

6) Find the values of each of the following:
   a) 7^{-2} $\frac{1}{49}$ b) 8^0 1   c) 2^{-10} $\frac{1}{1024}$

7) Rewrite each expression without using a negative exponent:
   a) x^{-5} $\frac{1}{x^5}$ b) $5x^3y^{-4}$ $\frac{1}{5x^3} \cdot \frac{1}{y^4}$ $\frac{1}{3x^4} \cdot \frac{5x^3}{1}$ ✗ $\frac{5x^3}{y^4}$ c) $\frac{3x^{-4}}{5x^3}$ $\frac{1}{3x^4}$ $\frac{3}{5x^7}$

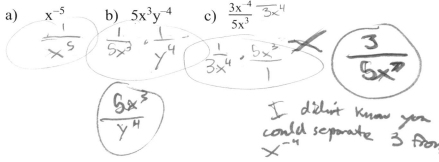

I didn't know you could separate 3 from x^{-4}

— Exponents & Polynomials —

Base Tables

Two's

N	2^N
-5	$1/32$
-4	$1/16$
-3	$1/8$
-2	$1/4$
-1	$1/2$
0	1
1	$2^1 = 2$
2	$2^2 = 4$
3	$2^3 = 8$
4	$2^4 = 16$
5	$2^5 = 32$
6	$2^6 = 64$
7	$2^7 = 128$
8	$2^8 = 256$
9	$2^9 = 512$
10	$2^{10} = 1024$

Three's

N	3^N
-5	$1/243$
-4	$1/81$
-3	$1/27$
-2	$1/9$
-1	$1/3$
0	1
1	$3^1 = 3$
2	$3^2 = 9$
3	$3^3 = 27$
4	$3^4 = 81$
5	$3^5 = 243$
6	$3^6 = 729$
7	$3^7 = 2187$
8	$3^8 = 6561$
9	$3^9 = 19683$
10	$3^{10} = 59049$

Five's

N	5^N
-5	$1/3125$
-4	$1/625$
-3	$1/125$
-2	$1/25$
-1	$1/5$
0	1
1	5
2	25
3	125
4	625
5	3125
6	15625
7	78125
8	390625
9	1953125
10	9765625

Ten's

N	10^N
-5	$1/100000$
-4	$1/10000$
-3	$1/1000$
-2	$1/100$
-1	$1/10$
0	1
1	10
2	100
3	1000
4	10000
5	100000
6	1000000
7	10000000
8	100000000
9	1000000000
10	10000000000

— Exponents & Polynomials —

Problem Set #9

Section A
Simplify.
1) $x^2 + y^2 - 3x^2 + 4y^3$
2) $xy + xy^2 + x^2y$
3) $(xy^2z^3)^4$
4) $x^5yz^4 + x^5yz^4$
5) $(x^5yz^4)(x^5yz^4)$
6) $5xy^2(2xy^3 - 6x^4y - 7x^5)$
7) $(4x^3z^6)(5x^5y^3)(2x^2y^4)$
8) $10x^2y^5(2x^4y^3z^2)^3$
9) $3x^4(9x - 2)(x - 5)$
10) $10x^3(x - 4)^2$
11) $(x - 10)^3$
12) $(x + 5)(x + 7)$
13) $(x - 3)(x + 6)$
14) $(x + 5)(x - 8)$
15) $(x - 1)(x - 12)$
16) $(x - 4)(x + 4)$
17) $(x^3 - 4)(x + 4)$
18) When multiplying two binomials, under what conditions does your final answer end up with…
    a) Four terms?
    b) Two terms (binomial)?
19) a) Simplify $(x - 3)^2$
    Evaluate both given $x = 7$
    b) $(x - 3)^2$
    c) $x^2 - 6x + 9$
    Evaluate both given $x = -2$
    d) $(x - 3)^2$
    e) $x^2 - 6x + 9$
    f) What do the above answers demonstrate?

20) **Simplify.** Assume that x is positive.
    a) $\sqrt{36x^{36}}$ b) $\sqrt{100x^{100}y^4}$
21) **Simplify.** Give answers without negative exponents.
    a) $(\tfrac{3}{4})^{-1}$ e) $6x^{-3}$
    b) 13^0 f) $\dfrac{6}{x^{-3}}$
    c) 40^{-2} g) $\dfrac{7x^{-4}}{x^5}$
    d) $(\tfrac{2}{3})^{-2}$ h) $\dfrac{12x^8y}{16x^2y^5}$

Section B
Simplify, and give your answers in two forms:
a) With denominators, but without negative exponents.
b) With negative exponents but without denominators.

Example: $\dfrac{4x^3}{x^8}$

Solutions:
a) $\dfrac{4}{x^5}$ b) $4x^{-5}$

22) $\dfrac{7x^2}{x^5}$ a)   b)
23) $\dfrac{6x^5y^7}{x^3y^{10}}$ a)   b)

Simplify.
24) $(x^5 - 6y)^2$
25) $(x^5 + 6y)(x^5 - 6y)$
26) $(x^5 + 6y)(x^8 - 6y)$
27) $(x^3 - 6y^2)(3x^3 + 2y^2)$
28) $(2x - 3)(x - 5)(x + 2)$
29) $(x + 4)^2(x - 1)^2$

— Exponents & Polynomials —

Problem Set #10

Section A
Simplify.
1) $-a^2 + b^2 + 5a^2 - 3b^2$
2) $a^3b^2 - b^2 - 5a^3b^2 + 5b^2$
3) $(a^4b^5)(a^5b^4)(5)$
4) $(a^3b^2)(-b^2)(-5a^3b^2)(5b^2)$
5) $(x + 3)(x - 5)$
6) $(x + 7)(x - 5)$
7) $(x - 5)(x - 3)$
8) $(x + 10)(x + 11)$
9) $(x + 10)^2$
10) $(x^4 - 7)(x^4 + 7)$
11) $(x^4 - 7)^2$
12) $x^3(x^4 - 7)(x^4 - 7)$
13) $(x^3 + 3y)(x^2 - 2y)$
14) $2p^2q^3(3pq^4)^2$
15) $2p^2q^3(3p + q^4)^2$
16) $3xy(x - 2y)(3x - 4y)$
17) $(x + 5)^3$

18) **Simplify.** Assume that x is positive.
   a) $\sqrt{16x^8}$ b) $\sqrt{25x^{16}y^{10}}$

19) **Simplify.** Give answers without negative exponents.
   a) $\left(\frac{5}{6}\right)^{-2}$ b) $(4x^2y^{-3})^{-2}$

20) a) Simplify
   $\frac{1}{4}x^2y(2xy^2)^3$
   Evaluate given x = 4; y = ½
   b) $\frac{1}{4}x^2y(2xy^2)^3$
   c) $2x^5y^7$

Evaluate given x = –2; y =1
   d) $\frac{1}{4}x^2y(2xy^2)^3$
   e) $2x^5y^7$
   f) What do answers b through e demonstrate?

21) **Rewrite** each number (given in scientific notation) in three ways (which are all equal to the original number):
   • Without a decimal point.
   • Without a negative exponent.
   • Standard decimal form.

   Example: $6.38 \cdot 10^{-5}$
   Solutions: $638 \cdot 10^{-7}$; $\frac{6.38}{10^5}$;
   and 0.0000638

   a) $7.6 \cdot 10^{-3}$
   b) $5.107 \cdot 10^{-2}$

Convert into scientific notation:
22) 0.0000064
23) 45,300,000,000
24) 0.0006002
25) 8,700

Convert into standard decimal form:
26) $9.2 \cdot 10^8$
27) $7.39 \cdot 10^{-5}$
28) $8.5472 \cdot 10^4$
29) $6.36 \cdot 10^{-1}$
30) $2.64 \cdot 10^0$

— Exponents & Polynomials —

Section B
Simplify.
31) $3xy(x-2y)(3x-4y)$
32) $(x-2y)(3xy)(3x-4y)$
33) $(x+3)(x-5)(x+5)$
34) $(x+3)(x-5)^2$
35) $(x+3)^4$
36) $\dfrac{5x^{-1}y^3z^{-2}}{3y^{-2}}$
37) $\dfrac{(x^8y^2z^{-2})^{-2}}{(x^{-4}z^3y^4)^5}$

Problem Set #11

Section A
Simplify.
1) $7x^3 - x^3$
2) $4x^3y^2 + 3x^3y$
3) $(4x^3y^2)(3x^3y)$
4) $\sqrt{25x^{100}}$
5) $\left(\frac{3}{4}\right)^{-1}$
6) $\left(\frac{4}{5}\right)^{-2}$
7) $(5x^{-3})^{-2}$
8) $\left(\frac{2x^{-3}}{3}\right)^{-3}$
9) $(x+6)^2$
10) $(x+6)(x-6)$
11) $(x+3)(x-4)$
12) $(x+17)(x-1)$
13) $(x-1)(x-12)$
14) $(3x-4)(2x+5)$
15) $(x+3y)(x-4)$

Solve for x in terms of y.
16) $y = 5x - 3$

Solve.
17) $4x + 9 = 5x - 2$
18) $8(4x+2) = -3(3x-6)$
19) $\dfrac{8}{3x-6} = \dfrac{-3}{4x+2}$
20) $7(x+3) = 12 - (2-x)$
21) $6 - 4(5x-3) + 4x$
    $= 5 - 2(x+3) + 5$
22) $(x+3)(x-2) = (x+9)^2$

Section B
Simplify.
23) $\dfrac{8x^3y^{-2}}{6x^{-5}z^{-4}}$
24) $\dfrac{4x^{-4}y^8z^{-7}}{20x^2y^3z^{-2}}$
25) $(x^3 - 2y^2)(x^5 + 4y^2)$
26) $(2x^2y)(3x^2y^3)(x^2y^3)$
27) $(2x^2y)(3x^2 - y^3)(x^2 - y^3)$
28) $(x-4)(x+2)^2$
29) $(x-4)(x+2)(x-2)$
30) $(3x-2)^3$

Solve for x in terms of y.
31) $\frac{1}{3}y - \frac{3}{4}x = \frac{2}{3}$

Solve.
32) $x(x+3)^2 = (x+2)^3$
33) $(6x-2)(3x-1) = (9x+1)(2x-3)$
34) $\frac{5}{8} - \frac{3}{8}\left(\frac{4}{9}x + 1\frac{1}{3}\right) = 2\frac{1}{4}\left(\frac{1}{3}x - \frac{8}{15}\right) - \frac{1}{2}\left(\frac{2}{3}x - 1\frac{2}{5}\right)$

Factoring

Problem Set #1

Group Work

In mathematics, factors of a given number are those numbers that go into it evenly. We can also *factor* a number into a product of its factors. For example: $35 \rightarrow 5 \cdot 7$

In algebra, we can also factor polynomials. For example, we can multiply $3x^4(5x + 6) \rightarrow 15x^5 + 18x^4$, or do the reverse and factor $15x^5 + 18x^4 \rightarrow 3x^4(5x + 6)$.

Multiply.
1) $7(4x - 3)$
2) $x^3(x^2 - 5)$
3) $3x^2(2x^3 + 7)$
4) $5x^2y^3(4x - 3y)$

Factor.
5) $28x - 21$
6) $x^5 - 5x^3$
7) $6x^5 + 21x^2$
8) $20x^3y^3 - 15x^2y^4$
9) $15x^4 + 25$
10) $7x^8 + 10x^5$
11) $12x^6 - 8x^5 + 20x^4$

Homework

Section A

Multiply.
12) $7(2x + 5)$
13) $5x^2(3x - 4)$
14) $6y^4(5y^3 + 3)$
15) $x^6(x^2 - 3x + 11)$

Factor. (Then multiply, in order to check your answer.)
16) $14x + 35$
17) $15x^3 - 20x^2$
18) $30y^7 + 18y^4$
19) $x^8 - 3x^7 + 11x^6$
20) $10x^4 - 15$
21) $y^8 - 4y^5$
22) $x^5 - 13x^4 + 6x^3$
23) $4x^7 + 12x^6 - 32x^5$

Section B

Multiply.
24) $(4x^2y^3)(2xy^3)$
25) $(4x^2y^3)(2x + y^3)$
26) $(4x^2 + y^3)(2x + y^3)$
27) $(5x^3y^4)^2$
28) $(5x^3 + y^4)^2$
29) $(5x^3 - y^4)(5x^3 + y^4)$
30) $(2x^3 - y^4)(5x^3 + y^4)$

Factor. (Then multiply, in order to check your answer.)
31) $12x^3y^5 + 8x^4y^4$
32) $9x^4y - 3x^3y^2 + 6x^2y^3$
33) $x^5 - 2x^2$
34) $10x^5 - 2x^2$

Note: The factoring that appears on this sheet is called *factoring out the GCF (greatest common factor)*.

— Factoring —

Problem Set #2

Group Work

On the previous set, the theme was factoring out the greatest common factor. In this set, we will be *factoring trinomials into the product of two binomials*. For example, we can either multiply:

$(x+4)(x+7) \to x^2 + 11x + 28$

or we can do the reverse and factor:

$x^2 + 11x + 28 \to (x+4)(x+7)$

Multiply.
1) $(x + 3)(x + 8)$
2) $(x + 10)(x + 2)$
3) $(x^4 + 5)(x^4 + 3)$

Factor.
4) $x^2 + 11x + 24$
5) $x^2 + 12x + 20$
6) $x^8 + 8x^4 + 15$
7) $x^2 + 9x + 14$
8) $x^2 + 13x + 42$
9) $x^2 + 16x + 48$
10) $x^6 + 6x^3 + 8$

Homework
Section A

Multiply.
11) $(x + 8)(x + 5)$
12) $(x + 3)(x + 4)$
13) $(x^5 + 3)(x^5 + 4)$
14) $(x + 1)(x + 6)$
15) $(x + 2y)(x + 5y)$
16) $(xy^2)(5xy^2)$
17) $(xy^2)(5x + y^2)$
18) $(x + y^2)(5x + y^2)$

Factor the trinomial. (As always, mentally multiply in order to check your answer.)
19) $x^2 + 13x + 40$
20) $x^2 + 7x + 12$
21) $x^{10} + 7x^5 + 12$
22) $x^2 + 7x + 6$
23) $x^2 + 7xy + 10y^2$
24) $x^2 + 10x + 21$
25) $x^2 + 14x + 24$
26) $x^2 + 10x + 24$
27) $x^2 + 25x + 24$

Factor out the GCF.
28) $7x^2 + 35$
29) $x^9 + 5x^7 - 17x^3$
30) $4y^3 + 6x^2 - 14$
31) $12x^6 + 16x^2$
32) $21x^3y^7 - 35x^6y^2$
33) $6x^5 + 21x^3 - 9x^2$

Mixed Factoring.
34) $x^7 - 5x^4$
35) $x^2 + 13x + 42$
36) $18x^5 + 30$
37) $4x^2 + 6x^5$
38) $x^2 + 10x + 16$
39) $2x^2 + 10x + 16$

Section B

Factor completely.
40) $6x^7 + 42x^6 + 72x^5$
41) $x^2 + 9xy + 14y^2$
42) $x^{12} + 9x^6 + 14$
43) $x^{12} + 9x^6y^2 + 14y^4$
44) $7x^{12} + 63x^6 + 98$
45) $8x^{13} + 72x^7y^2 + 112x$

— Factoring —

Problem Set #3

Group Work
Factor.
1) $x^2 + 9x + 20$
2) $x^2 + 14x + 45$
3) $x^2 - 14x + 45$
4) $x^2 + 8x - 20$
5) $x^2 - 8x - 20$
6) $x^2 + 15x + 54$
7) $x^2 - 15x + 54$
9. 8) $x^2 + 15x - 54$ ✗ $(x+18)(x-3)$
9) $x^2 - 15x - 54$
10) Use the above problems to formulate *Rules for Factoring a Trinomial.*

Factor.
11) $x^2 + 5x + 6$
12) $x^2 - 5x + 6$
13) $x^2 + 5x - 6$
14) $x^2 - 5x - 6$
15) $x^2 + 13x + 36$
16) $x^2 + 16x - 36$
17) $x^2 - 23x + 60$
18) $x^2 - 28x - 60$

Homework
Section A
Multiply.
19) $(x + 7)(x + 5)$
20) $(x - 7)(x + 5)$
21) $(x + 7)(x - 5)$
22) $(x - 7)(x - 5)$
23) $(x + 11)(x + 5)$
24) $(x - 11)(x - 5)$
25) $(x + 11)(x - 5)$
26) $(x - 11)(x + 5)$
27) $(x + 8y)(x - 10y)$
28) $(x^3 - 5)^2$

Factor.
29) $x^2 + 13x + 22$
30) $x^2 - 13x + 22$
31) $x^2 + 21x - 100$
32) $x^2 - 2x - 48$
33) $x^2 + 18x - 40$
34) $x^2 - 27x + 50$
35) $x^2 - 12x + 36$
36) $x^2 + 15x - 50$
37) $x^2 - x - 20$
38) $x^2 - 10x + 24$
39) $x^2 - 10x - 24$
40) $x^2 + 10x + 24$
41) $x^2 + 10x - 24$

Factor out the GCF.
42) $8x^2 + 16x - 40$
43) $12x^3 + 24x$
44) $8x^4y - 6x^2y^5 + 4x^4y^4$

Mixed Factoring.
45) $7x^5 + 3x^3$
46) $x^2 - 15x + 50$
47) $5x^2 - 15x + 50$
48) $2x^5 + 4x^4 - 16x^3$

Section B
Multiply.
49) $(x^3 + 5y^2)(x^3 - 2y^2)$
50) $(5x^3 + 13y)^2$

Factor Completely.
51) $x^2 - 21x + 90$
52) $x^7 + 21x^6 - 100x^5$
53) $x^{10} + 3x^5 - 10$
54) $17x^3 + 34x^2 + 17x$

— Factoring —

Problem Set #4

Group Work
Some of these are two-step factoring problems. Always try the GCF first, then try to factor the trinomial.

1) $4x^2 + 20x + 24$
2) $4x^2 + 20xy + 24y^2$
3) $x^2 + 4xy - 32y^2$
4) $10x^6 - 30x^5 - 180x^4$
5) $x^2 - 18x + 81$
6) $4x^2 + 24x + 36$
7) $4x^2 + 24x - 36$
8) $x^2y^5 + xy^5 - 56y^5$
9) $x^5 + x^4 - 20x^3$
10) $x^2 + 6x - 8$
11) $x^2 + 41x + 180$
12) $x^2 - 41x + 180$
13) $x^2 + 41x - 180$
14) $x^2 - 41x - 180$

23) $3x^2(x + 5)(x - 8)$
24) $5x(x - 3)^2$

Factor.
25) $x^2 + 7x + 12$
26) $x^2 - 7x + 12$
27) $x^2 + x - 12$
28) $x^2 - x - 12$
29) $x^2 + 10x + 21$
30) $x^2 + 10xy + 21y^2$
31) $x^8 + 10x^4 + 21$
32) $x^4 + 10x^2y^2 + 21y^4$
33) $21x^4y^3 - 28x^2y^2$
34) $x^2 + 13x + 30$
35) $x^2 + 13x - 30$
36) $x^2 - 13x + 30$
37) $x^2 - 13x - 30$
38) $4x^2 + 16x - 48$

Homework
Section A
Multiply.
15) $(x + 3)(x - 4)$
16) $(x - 3)(x + 4)$
17) $(x - 3)(x - 4)$
18) $(x + 3)(x + 4)$
19) $7x^4y^3(3x^3 + 4y^2)$
20) $3xy^4(x^2 + 2x^2y^2 - 5y^2)$
21) $(3x^2y^8)^2$
22) $4(x - 2)(x + 6)$

Section B
Factor Completely.
39) $4x^2y^3 + 20xy^4 + 24y^5$
40) $x^6 + 12x^3 + 32$
41) $2x^3y^2 + 8x^2y^3 - 64xy^4$
42) $3x^4y^4 - 30x^3y^4 - 33x^2y^4$
43) $7x^7 - 91x^6 + 210x^5$
44) $3x^4 - 9x^3 - 120x^2$
45) $5x^4y - 30x^3y + 45x^2y$
46) $x^2y^2 + 8xy + 15$
47) $22x^{20} + 220x^{10} - 858$

— Factoring —

Problem Set #5

Group Work

In this set of problems, you will practice factoring *the difference of two squares.* For example, when we multiply (x+4)(x−4) the middle terms cancel, so instead of getting a trinomial as an answer, we get the binomial x^2-16.

Going in reverse, we factor x^2-16 to get (x+4)(x−4).

Multiply.
1) $(x + 6)(x - 6)$
2) $(y^3 + 4)(y^3 - 4)$

Factor. (Hint: Some can't be factored.)
3) $x^2 - 36$
4) $y^6 - 16$
5) $x^8 - 9$
6) $x^2 + 25$
7) $x^9 - 25$
8) $x^8 - 8$
9) $x^{16} - 25y^2$
10) $9x^6 - 4y^{10}$
11) $3x^7 - 12x^3$
12) $x^{16} - 16$
13) What makes it possible to factor a binomial into the product of two binomials?

Homework
Section A
Multiply.
14) $(x + 3)(x + 5)$
15) $(x - 3)(x + 5)$
16) $(x - 3)(x - 5)$
17) $(x - 7)(x + 7)$
18) $(x + 7)^2$
19) $2x^5(x + 5)^2$

Factor the binomial.
20) $x^2 - 25$
21) $x^2 + 25$
22) $x^2 - 64$
23) $x^2 + 64$
24) $x^{25} - 25$
25) $x^{10} - 9$
26) $x^{10} - 12$

Mixed Factoring.
27) $x^2 - 10x + 25$
28) $x^6 - 8x^3 + 16$
29) $x^2 + 7x - 18$
30) $x^2 - 1$
31) $x^8 - 49$
32) $12x^2 - 12$
33) $3x^2 + 21x - 54$
34) $x^2 + 9x - 20$
35) $x^2 + x - 72$
36) $x^2y^2 + xy - 72$

Section B
Factor completely.
37) $x^{12} - 16$
38) $5x^5 - 20x^3$
39) $x^6 - 9y^4$
40) $6x^3y^6 + 14x^5z^3$
41) $2x^3 - 26x^2 + 24x$
42) $2x^5 + 38x^4 + 96x^3$

— Factoring —

Problem Set #6

Review!
Section A
Multiply.
1) $(x + 9)(x - 9)$
2) $(x - 9)^2$
3) $(x^2 - 6)(x^2 + 6)$
4) $(x^2 - 6)^2$
5) $(x - 15)(x + 15)$
6) $(x - 18)(x + 2)$
7) $(x + 30)(x - 30)$
8) $(4x^2)(5xy^3)(5xy^3)$

Factor.
9) $x^2 - 900$
10) $x^3 - 9x$
11) $x^6 - y^4$
12) $x^2 + 25x + 84$
13) $x^2 - 25x - 84$
14) $x^2 + 25x - 84$
15) $x^2 - 25x + 84$

Simplify.
16) $\sqrt{49x^{16}}$
17) $\left(\frac{2}{7}\right)^{-2}$
18) $\left(3x^{-2}\right)^{-4}$
19) $\frac{3x^{-2}}{9x^{-6}}$

Evaluate
20) given $x = 5$ and $y = -\frac{1}{8}$
    Evaluate $x^2 - 64y + y^2 + \frac{1}{x}$

Solve for x in terms of y.
21) $y = 4x + 9$

Solve.
22) $5(x + 2) - 4(7 - x) = 6$
23) $\frac{2}{x+4} = \frac{3}{x-1}$
24) $(x + 1)(x - 2) = (x + 3)^2$

Section B
Simplify.
25) $(x - 3)(x - 4)^2$
26) $(x - 6)^3$
27) $(x^2 + 9)(x + 3)(x - 3)$
28) $(4x^2)(5x - y^3)(5x + y^3)$
29) $\left(\frac{3y^{-3}}{2x^3}\right)^{-2}$
30) $\frac{5x^{-4}y^{-3}}{15x^{-3}y^5}$

Evaluate
given $x = -\frac{1}{2}$ and $y = \frac{3}{4}$
31) $\frac{y}{3x} - x^3 + 6y$

Factor.
32) $x^6 - x^4$
33) $2x^9 - 18x^3y^4$
34) $x^2 + 25x - 70$
35) $5x^2 + 25x - 70$
36) $x^6 - 18x^3 + 81$
37) $x^6 - 81$
38) $x^{12} - 81$
39) $x^4 - 3x^3 - 28x^2$

Solve for x in terms of y.
40) $y = \frac{2}{3}x - 2$

Solve.
41) $-\frac{3}{5}X - 1 = \frac{5}{6} - 2X$
42) $\frac{-\frac{1}{2}}{3X + 1\frac{1}{4}} = \frac{4}{\frac{3}{4}X - \frac{1}{2}}$

— Factoring —

Problem Set #7

Group Work

You will now practice factoring trinomials that have a number before x^2.

For example, try factoring $42x^2 + 41x + 9$. It can be helpful to write the pairs of factors of 42 (which are 1,42; 2,21; 3,14; & 6,7) and 9 (1,9; 3,3). The answer turns out to be $(14x + 9)(3x + 1)$.

Factor.
1) $7x^2 + 30x + 8$
2) $7x^2 - 15x + 8$
3) $7x^2 - x - 8$
4) $7x^2 + 10x - 8$
5) $7x^2 - 55x - 8$
6) $15x^2 - 28x + 12$
7) $15x^2 - 41x - 12$
8) $6x^2 + 17x - 45$
9) $12x^2 + 36x + 24$
10) $3x^5 - 8x^4 + 4x^3$

Homework

Section A

Multiply.
11) $(x + 8)(x - 8)$
12) $(x - 8)^2$
13) $(x + 15)(x - 15)$
14) $(x + 15)^2$

Factor.
(Hint: Each one is possible.)
15) $8x^2 + 22x + 15$
16) $8x^2 + 34x + 15$
17) $8x^2 - 14x - 15$
18) $6x^2 - 5x - 4$

Review

Factor.
19) $x^2 + 8x + 16$
20) $x^2 - 8x + 16$
21) $x^2 - 8x + 15$
22) $x^2 - 8x - 2$
23) $x^2 - 8x + 2$
24) $x^6 - 400$
25) $x^6 - 40$
26) $x^5 - 25$
27) $x^4 + 25$
28) $x^4 - 81$
29) $x^4 - 18$
30) $2x^4 - 18$
31) $9x^4 - 25y^6$
32) $x^7 - 16x^3$

Section B

Factor.
33) $x^2 + xy - 90y^2$
34) $3x^2 + 33x - 36$
35) $x^8 - x^4 - 30$
36) $x^8 - 9x^7 + 18x^6$
37) $x^2 + 30x + 216$
38) $x^2 - 30x - 216$
39) $8x^2 - 62x + 15$
40) $8x^2 + 29x + 15$
41) $8x^2 - 29x + 15$
42) $8x^2 + 19x - 15$
43) $8x^2 + 37x - 15$
44) $8x^2 - 37xy^3 - 15y^6$
45) $6x^6 + 10x^3y^2 - 4y^4$

— Factoring —
Problem Set #8

Group Work
We will now begin solving *quadratic equations*, which are equations with an x^2 term.

The key to this is realizing that if the product of two terms equals zero, then one of the two terms must be equal to zero. In other words, if $a \cdot b = 0$, then either a or b must be zero. Likewise, if $(x-3)(x+5) = 0$ then x must be either equal to 3 or −5. (It is important that you understand this last statement!)

Here's how this can be used to solve a quadratic equation:
Example:
Solve $x^2 + 3x = 8x - 6$
$x^2 + 3x - 8x + 6 = 0$
$x^2 - 5x + 6 = 0$
$(x - 3)(x - 2) = 0$
x = 2 or 3

Note that we can check our answer by plugging either 2 or 3 into the original equation and it should work!

Solve.
1) $x^2 - 7x = 6x - 30$
2) $x^2 + 13x = 10x + 28$
3) $4x^2 + 13x = 3x^2 - 30$
4) $4x^2 + 13x = 5x^2 - 30$
5) $x^2 + 24 = 2x(x - 5)$

Homework
Section A
Multiply.
6) $(x + 6)(x - 3)$
7) $(2x + 6)(x - 3)$
8) $4(x + 8)(x - 12)$
9) $(x + 8)4(x - 12)$
10) $(4x + 32)(x - 12)$
11) $(x + 8)(4x - 48)$
12) $(x^2 - 5y)^3$

Factor.
13) $x^8y^4 - 9z^6$
14) $y^4 + 9z^6$
15) $x^2 + 15x + 54$
16) $x^2 + 29x + 54$
17) $x^2 - 15x + 50$
18) $x^2 - 100$
19) $18x^4y^3 + 12x^2y^7$
20) $x^8 - 1$
21) $14x^2 - 29x + 12$

Solve.
22) $x^2 = 10x - 16$
23) $7x + 18 = x^2$
24) $x^2 + 18x + 80 = 0$
25) $x^2 - 15 = 14x$
26) $6x^2 - 90 = 5x^2 - 9x$
27) $x^2 + 3x = 6x + 4$

Section B
Factor.
28) $14x^{10} - 29x^5 + 12$
29) $14x^2 - 29xy + 12y^2$
30) $14x^4 - 29x^2y + 12y^2$
31) $10x^5y^2 + 20x^4y^4 - 350x^3y^6$
32) $14x^2 - 22x - 12$
33) $x^3 - 21x^2 + 20x$

Solve.
34) $x^2 + 3x + 24 = 3x^2 + x$
35) $(x + 5)(x + 3) = -1$

— Factoring —
Problem Set #9

Section A
Factor.
1) $x^2 - x - 20$
2) $x^2 + 6x - 36$
3) $18x^2 + 31x + 6$
4) $14x^2 + 13x - 12$
5) $x^2 - 225$
6) $x^2 + 225$
7) $x^2 + 9x - 20$
8) $5x^5 + 20x^3$

Multiply.
9) $(x^3 - 6)(x^3 + 6)$
10) $(x^3 + 6)^2$
11) $(x - 40)^2$

Solve.
12) $x^2 - 7x - 30 = 0$
13) $x^2 + 25 = 10x$
14) $7 + 2x = 8x + x^2$
15) $x^2 + 5x = 6$
16) $x^2 + 5x + 6 = 0$
17) $x^2 + 5x + 6 = 2$
18) $x^2 + 5x + 6 = -2x$
19) $x^2 + 5x + 6 = 2x^2$
20) $x^2 - 54 = 25x$
21) $2x^2 - 108 = 50x$
22) $4(3x - 2) = 12x - 8$

Section B
Multiply.
23) $4x^3(x + 3)(x - 3)$
24) $(3x - 4y^3)^2$
25) $(x^{10} + 100)(x^5 + 10)(x^5 - 10)$

Factor.
26) $x^9 - x$
27) $12x^3y^5 - 4x^2y^3$
28) $10x^3 + 10x^2 - 200x$
29) $8x^2y^5 + 24x^5y^2$
30) $8x^2y^5 + 24x^2y^5$
31) $x^{12} - 625$
32) $18x^2 + 12x - 6$
33) $18x^2 - 21xy + 6y^2$
34) $18x^6 - 107x^3 - 6$

Solve.
35) $13 - (x + 3)^2 = 12$
36) $5x^2 - 8x + 3 = x^2 + 12x - 21$
37) $7x^2 + 3 = x^2 + 19x - 12$
38) $x^2 - 56 = (x + 2)(x - 8)$
39) $2x^2 - 56 = (x + 2)(x - 8)$

— Factoring —

Problem Set #10

Section A
Factor.
1) $x^2 - 27x + 50$
2) $x^2 + 17x - 60$
3) $x^6 - 100$
4) $x^8 - 10{,}000$

Solve.
5) $x^2 + 4x + 4 = 0$
6) $x^2 + 4 = -4x$
7) $x^2 = 4x + 45$
8) $7x = x^2 + 10$
9) $x^2 = 13x - 12$
10) $14 - 7(x+3) = x^2 + 3$
11) $7x - 7 = x^2 + 5$
12) $7x - 7 = x + 5$
13) $\frac{6}{x+13} = \frac{x}{x+3}$
14) $5x = (x-8)(x+3)$

Section B
Factor.
15) $x^2 - 20x + 91$
16) $20x^2 - 48x + 16$
17) $20x^2 + 321xy + 16y^2$
18) $x^{16} - 1$
19) $20x^2 + 59x - 16$
20) $20x^8 - 32x^4y^3 - 16y^6$
21) $3x^6 - 15x^5 + 6x^4$

Solve.
22) $5x(x-3) = 4x^2 - 50$
23) $x^2 - 7x - 10x = 36 - 17x$
24) $(x+3)^2 = 15 - (3-4x)$
25) $(x-3)^2 = (x+5)(x-5)$
26) $(x-5)(x-7) = 4x^2 + 6x - 85$
27) $300 - 3x^2(x-4) = (x^2 - 100)(2x - 3)$

Problem Set #11

Section A - Solve.
1) $x^2 + 77 = 18x$
2) $(x-4)(x-10) = 55$
3) $(x-4)(x-7) = 0$
4) $3x^2 + 5 = (x+7)^2 - 8$
5) $(x-4)^2 = 7x^2 - x + 13$
6) $x^2 + 4x = 4x + 64$
7) $\frac{12}{x+6} = \frac{4}{3x+2}$
8) $6x^2 - 9x = 5x^2 + 2x - 24$
9) $x^2 + 6x = 3$

Section B - Solve.
10) $(2x-3)(x+8) = 60$
11) $7x^3 = 10x^3 - 300x$
12) $\frac{x-2}{2x-25} = \frac{3}{x+20}$
13) $6x^5 - 7x^4 = 11x^4 - 12x^3$
14) $(x+3)(3x-5) = 3x^2 + 4x - 15$
15) $7x - 5 = x(x+7) - 105$
16) $7x - 5 = 7(x+7) - 105$
17) $2x^2 - 5x = 3x^2 - x - 60$
18) $2x^2 - 5x = 2x^2 - x - 60$

19) $-12x + 3x^2 + 2x - 7 = 13x^2 - 20x - 8x^2 + 12x - 47$
20) $5x^2 + 3x^3(x-3) = 5x^2(3x-8)$

— Factoring —

Problem Set #12

Group Work

Word Problems

(A segue way into the next unit.)

   Algebra is the language of mathematics. Its power comes from its ability to succinctly express mathematical concepts that in English would be lengthy or awkward. Often, the challenge of a word problem is found in translating thoughts, which are expressed in English, into algebraic expressions and equations.

Translate into English.
Example: $3x + 5$
Solution: Five more than three times a number.
1) $2x + 3$
2) $3x - 8$

Translate into Algebra.
3)   Four less than five times a number.
4)   The square of one more than a number.

Find the Number.
5)   Three less than twice a number is eight.

Homework

Solve.
6) $x^2 - 5x = 2x + 25 - 7x$
7) $(x-4)^2 = x^2 + 16$
8) $5x = x^2 - 24$
9) $(x+4)(x-3) = 18$
10) $120 - 60x^2 = 420 - 490x$
11) $x - 5 = \frac{7x}{x+4}$
12) $\frac{5}{x-4} = \frac{x+2}{8}$
13) $\frac{5}{x-4} = \frac{8}{x+2}$
14) $(x-7)^2 = 4x^2 + 7x + 79$
15) $5x^3 - 20x^2 = 3x^3 + 48x$
16) $2x^2 - 11x = (x+8)(x-5)$
17) $x^2 - 11x = (x+8)(x-5)$

Translate into English.
18) $x^2 + 10$
19) $6x - 1$

Translate into Algebra.
20)   13 more than twice a number.
21)   Five less than half a number.

Find the Number.
22)   Three more than twice a number is 24.

23)   Solve: $2x^3(x-4)(3x+5) - (6x)(2x^2) = 2x^5 + 34x^4$

Word Problems

Problem Set #1

Group Work
Relating numbers.
If we are given two numbers, then we can make statements (in either English or algebra) that relate the numbers.
Example: 6 and 10
Possible statements:
- One number is four greater than the other; $x = y + 4$
- One number is $3/5$ of the other; $y = 3/5\, x$
- The difference of the two numbers is four; $x - y = 4$
- The larger number is eight less than three times the smaller; $x = 3y - 8$

Give at least five statements (in both English and algebra) that relate each pair of numbers.
1) 7 and 3
2) 4½ and 9
3) 8 and 13

Homework
Translate into Algebra.
4) Seven more than twice a number.
5) One number is two more than five times another.
6) Six less than half a number.
7) Half of six less than a number.
8) The sum of two numbers.
9) The sum of two numbers is 18.
10) The product of two numbers is 18.
11) The square of three less than a number.

Translate into English.
(Try to avoid using the words "plus", "minus", "equals", "x", "y", etc.)
12) $6x - 3$
13) $6(x - 3)$
14) $x^2 + 5$
15) $x^2 + y^2$
16) $(x+y)^2$
17) $y = x^2 + 5$
18) $4x - 1 = 5$
19) $x + y = 7$
20) $y - x = 7$

Find the Number.
21) Four more than three times a number is 22.
22) Eight less than three times a number is six.
23) Half a number is twelve less than twice that number.
24) The square of a number is 14 more than five times that number.

— Word Problems —

Problem Set #2

Group Work
Find the Number.

1) Eight more than three times a number is four.
2) One-half of three less than a number is six.
3) Four less than eight times a number is 37.
4) Consider these statements:
   - The sum of two numbers is 13.
   - The larger number is three greater than the smaller.
   - The larger of two numbers is one more than three times the smaller.

   a) This statement: "The smaller number is three less than the larger" is equivalent to which of the above statements?
   b) How many possible solutions are there to the first statement alone?
   c) How many possible solutions are there to the second statement alone?
   d) How many possible solutions are there that satisfy both the first and second statements?
   e) How many possible solutions are there that satisfy both the first and third statements?
   f) How many possible solutions are there that satisfy both the second and third statements?
   g) How many possible solutions are there that satisfy all three statements?

Two-number Riddles

5) Solve this riddle: The smaller of two numbers is four less than the larger. The larger is one less than twice the smaller.
6) Make your own two-number riddle! Start by choosing two numbers (between 1 and 20). Make two statements about your two numbers.

Homework
The homework problems should be selected from the two-number riddles that the class has made up!

— Word Problems —

Problem Set #3

Group Work

1) With the equation $y = 5x - 7$.
   a) Find y when x is 4.
   b) Find y when x is $-3\frac{1}{2}$.
   c) Find x when y is 4.
2) With the equation $4x + 3y = 12$.
   a) Find x when y is 2.
   b) Find x when y is -7.
   c) Find y when x is $-\frac{3}{4}$.
3) Give three solutions to $y = 2x - 3$.
4) Give three solutions to $5x - 3y = 4$.
5) Find a solution that works for both $y = 2x - 3$ and $5x - 3y = 4$.
6) *Challenge Problem!*
   Here are four statements:
   1. The sum of two numbers is seven.
   2. The larger number is twice the smaller.
   3. Three times the larger number is 35 greater than four times the smaller.
   4. The larger number is one more than the square of the smaller.

   How many two-number riddles can you create by selecting any two of the above statements? Solve each one!

Homework

7) Eight more than ten times a number is 120. Find the number.
8) The square of a number is 21 less than ten times that number. Find the number.
9) In a basketball game, the Tigers beat the Apes by 18 points. Twice the Tigers' score was six less than three times the Apes' score. What was the Tigers' score?
10) Given $y = \frac{2}{3}x + 4$
    a) Find y when x is 6.
    b) Find x when y is -7.
    c) Find y when x is $-\frac{3}{4}$.
11) Give three solutions to $x + 2y = 7$.

***Three two-number riddles, written by students, should be added here!

— Word Problems —

Problem Set #4

Group Work
Find the numbers.

1) The sum of two numbers is 17 and the sum of their squares is 185.

2) The difference of two numbers is 16. Four times the smaller number is 13 less than three times the larger number. What are the numbers?

3) The sum of two consecutive integers is 31.

4) The sum of two odd consecutive integers is 48.

5) The sum of two even consecutive integers is 34.

6) Find the common solution:
$y = 2x + 7$
$3x + 4y = 6$

Homework
Section A
Find the numbers.

7) The sum of two numbers is 210 and their difference is 40.

8) Two consecutive integers are such that four times the smaller is four more than 3 times the larger.

9) The product of two numbers is 80 and one number is one more than three times the other.

Find the common solution to each pair of equations.

10) $y = x + 2$
$y = 2x - 1$

11) $2x + y = 5$
$x + y = 4$

12) $5x + 3y = 1$
$x - 3y = 9$

Section B
Find the numbers.

13) The sum of two numbers is 335. The larger number is 40 less than twice the smaller.

14) Together a coffee and a donut cost $3.35. The donut costs 40¢ less than twice the price of the coffee. Find the price of the donut.

Find the common solution to each pair of equations.

15) $y = 2x + 4$
$3y - 5x = 9$

16) $x = 4y + 1$
$3y + 2x = 7$

17) $5x - y = 3$
$2y - x = 12$

— Word Problems —

Problem Set #5

Group Work

Find the common solution.
1) $4x + y = 1$
   $2x + 2y = 5$
2) $3x - 5y = 5$
   $2x + 3y = 16$
3) Give two solutions to $3x - 7y = 21$

Find the test average.

Note: In this unit, we will use the following percent equivalents for grades:
A+ = 98%; A = 95%; A- = 92%,
B+ = 88%; B = 85%; B- = 82%, Etc.

4) The first test is 17 out of 20, the second test is 18 out of 25, and both tests have equal weight.
5) The first test is 17 out of 20 and is worth 25%, and the second test is 18 out of 25 and is worth 75%.
6) The first test is 17 out of 20 and is worth 35%, and the second test is 18 out of 25 and is worth 65%.
7) The first test is a C and is worth 25%, the second test is a D and is worth 15%, and the third test is a B+ and is worth 60%.
8) Bill's age is a year less than twice Jane's age. Five years ago, Bill's age was three times Jane's age. Find Bill's age.

Homework

Section A

9) Twice a smaller number is 18 less than the larger, and their difference is 11. What are the two numbers?
10) The difference of two numbers is five, and the sum of their squares is 233. Find the two numbers.
11) Fran has two dollars less than twice as much money as Mary. How much does Mary have if they have $41.50 together?
12) Joe received a 95% on the final exam which was worth 60% of the class grade. If he received a 67% on the midterm exam, what was his final grade in the class if the midterm exam was worth 40% of the class grade?
13) Find the common solution to each pair of equations:
    a) $x + y = 1$
       $x - y = 6$
    b) $4y - 3x = 12$
       $3x + 11y = -7$
    c) $3x + 2y = 5$
       $3x + 5y = 32$
14) Give two other equations that have all the same solutions as $2x - 3y = 7$

Section B

15) The sum of the squares of two consecutive odd integers is 394.

16) Emily received a C and a B on her two essays, which were each worth 20%, an A– for class participation (worth 15%), and a D on her final exam (worth 45%). What is her final course (letter) grade?

17) Find the common solution to each pair of equations:
   a) $y = 3x - 1$
      $4x - 3y = 13$
   b) $x + 2y = 4$
      $3x - 4y = 17$
   c) $5x - 2y = 20$
      $2x + 7y = -5$

Problem Set #6

Group Work

1) Jeff has 20 coins in his pocket worth a total of $3.95. If he has only quarters and dimes, then how many of each type of coin does he have?

2) Bob has a handful of nickels and dimes worth $2.45. How many dimes are there if there are four more nickels than dimes?

3) In Kate's math class, the final exam is worth 30%. Her test average going into the final is 87.5. What does she need to score on the final exam (rounded to the nearest whole number) in order to end up with at least a 90 test average?

Homework

Section A
Find the common solution.

4) $8x + 3y = 12$
   $8x + 12y = -39$

5) $5x - 7y = 3$
   $x + 7y = 15$

6) $6x - 2y = 9$
   $2x + y = -2$

7) One number is three more than another. Four times the smaller number is seven more than three times the greater. Find the two numbers.

8) The difference of two numbers is 7, and the smaller number is 65% of the larger. Find the two numbers.

— Word Problems —

9) Twice Bill's weight is 12 kilograms less than Frank's weight, but three times Bill's weight is three kilogram's more than Frank's weight. Find Frank's weight.

10) Jim's math course's final grade is calculated with these weights: final exam 25%, quiz average 35%, homework 30%, and class participation 10%. If Jim receives scores of 77, 93, 65, and 80 on those four categories, respectively, then what is his final grade (rounded to the nearest whole number)?

11) Bill is two-thirds of Mark's age. If Mark is 5 years older than Bill, then how old is Bill?

Section B

12) Find common solution:
$2x - y = 6$
$x^2 + 4x - 3y = 26$

13) Jeff is half as old as Pete. Next year the sum of their ages will be 35. How old is Jeff?

14) Sue is 20 years younger than Gail. In 9 years, Sue will be $3/5$ as old as Gail. How old is Sue now?

Problem Set #7

Group Work

1) Tickets at a concert cost $8 for section A and $4.25 for section B. 4500 tickets were sold for a total of $30,000. How many tickets of each type of were sold?

2) Joe earns $10.50/hr at a restaurant and $8/hr at a movie theater. Last week, between the two jobs, he worked 17 hours and earned $159.75. How many hours did he work at each job?

Homework
Section A
Find the common solution.

3) $x + 8y = 17$
$5x + 8y = 3$

4) $4x + 2y = 10$
$2x - 3y = 1$

5) $3x + 5y = 3$
$2x - 3y = 5$

6) $y - 3x = 4$
$y - 3x = 7$

7) Hannah is 8 and her father is 30. How long will it be until Hannah is half her father's age?

8) It took Tim 17 minutes to drive 11 miles to get to the park. He then walked 2½ miles in 51 minutes, and lastly, ran 1½ miles in 12 minutes. What was his average speed for the whole trip?

9) Kate has a pocketful of dimes and quarters. How many quarters are there if there are a total of 29 coins and they are worth $4.70?

Section B

10) On Saturday, Ben jogged for 2½ hours. On Sunday, he jogged for two hours, but went 2 km further, and jogged at a rate that was 3 km/h faster than he did on Saturday. How far did he jog on Saturday?

11) Jeff biked for two hours at 4 mph and then biked for two hours at 18 mph. What was his average speed?

12) Mary biked up an 18-mile hill at 4 mph and came back down (the same route) at 18 mph. What was her average speed?

13) Max biked up a 3-mile hill at 4 mph and came back down at 18 mph. What was his average speed?

14) Margaret biked up a hill at 4 mph and came back down at 18 mph. What was her average speed?

15) There are two numbers. One number is 60 greater than the other. The greater number is one-quarter of the lesser. Find the numbers.

Problem Set #8

Group Work

1) A thief crosses a bridge at 9:37pm going 60 mph in a car. At 9:49pm, a police car chasing the thief and going 75 mph crosses the same bridge. Assuming that the cars maintain their speeds, at what time, and how far from the bridge, does the police car catch the thief?

Homework
Section A

2) Find the common solution:
  a) $3x + y = 5$
     $4x + 7y = 1$
  b) $-3x + 2y = 1$
     $-2x - 3y = -6$

3) The difference of two numbers is 5, but the sum of their squares is 157. Find the two numbers.

4) Four boxes of apples and six boxes of pears cost $90. Two boxes of apples and five boxes of pears cost $61. How much does one box of pears cost?

5) Sal drives up a 5 mile hill in 12 minutes. He then turns around and drives down the hill in 8 minutes. What was the average speed for his entire trip?

6) One number is one less than twice another number. Twice the sum of the numbers is 49. Find the two numbers.

— Word Problems —

Section B

7) Find two numbers such that they are in a ratio of 4:5 and their average is 18. Find the two numbers.

8) Bill is five years younger than Len. In two years, Len will be twice as old as Bill. How old is Bill now?

9) Maria has a handful of quarters, dimes, and nickels worth a total of $2.40. There are a total of 20 coins and 2½ times as many nickels as dimes. How many of each type of coin are there?

10) Thomas and Keith start out 12 miles apart. At what time do they pass each other if they both start biking toward each other at 2:20pm, and Thomas bikes at 15 mph and Keith bikes at 21 mph?

11) Find the common solution:
$3y - x = 6$
$6y = 2x + 12$

12) Give an equation that has both (3,1) and (5,4) as a solution.

Problem Set #9

Group Work

$C = \frac{5}{9} \cdot (F - 32)$

$F = \frac{9}{5} \cdot C + 32$

1) Use the above formulas to answer the following questions:
   a) Convert 95°F to °C
   b) Convert 10°C to °F
   c) Convert 43°F to °C
   d) Convert 43°C to °F

Graphing

2) Follow these instructions:
   a) Take a full-size sheet of graph paper and put in a "portrait" orientation.
   b) Draw the horizontal axis near the bottom of the page. Label it "Celsius", and have the range of the numbers on the axis go from 0 to 60.
   c) Draw the vertical axis near the left side of the page. Have the range of the numbers on the axis go from 30 to 150, and label it "Fahrenheit".
   d) Plot each of the four answers from problem #1 on the graph.

— Word Problems —

3) Use the graph you have just created to estimate the following questions:
   a) Convert 140°F to °C
   b) Convert 0°C to °F
   c) Convert 77°F to °C
   d) Convert 20°C to °F
   e) Convert 50°C to °F

Homework

Section A

4) Find the common solution:
$$4x + 3y = 1$$
$$-3x + 5y = -2$$

5) The sum of two numbers is 32. The larger number is 12 greater than twice the smaller. Find the numbers.

6) Find three consecutive integers whose sum is –63.

7) Find two numbers with a ratio 2:3 and average of 35.

8) Bill is four times as old as Clara. In four years he will be three times as old as her. How old is Clara now?

9) In the morning, Jane travels to work at an average speed of 45 mph. She takes the same route home but only averages 30 mph. If her total travel time is 1 hour and 15 minutes, how far is Jane's place of work from her home?

10) On a given flight, an airline offers two types of seats: first class for $500 and economy for $300. 110 tickets were sold for a total value of $38,400. How many first class seats were sold?

Section B

11) Keith is 25% as old as Bill. After how many years will Keith be 40% as old as Bill if Keith is 14 now?

12) A store is making a trail mix by mixing nuts (which cost $5.50/lb) and dried fruit (which cost $4.20/lb). At what ratio should they be mixed in order to get a trail mix worth $5.00 per pound?

13) At 1:15pm Sam left his house, biking at 12 mph. After getting a flat tire, he walked back home at 3 mph, arriving at 2:30pm. How far from home did his bike break down?

14) A train leaves Bigtown at 70 mph toward Smallville (545 miles away) at 1:20pm. At 1:50pm, another train leaves Smallville, heading for Bigtown, at 50 mph. At what time, and how far from Bigtown, do they pass one another?

15) Derive an average speed formula given R_1 is the rate of speed traveling between two points, and R_2 is the rate of speed returning back along the same route.

Midyear Review

Problem Set #1

Note: In this unit, a calculator is permitted for the sections on *percents* and *Proportions & Dimensional Analysis* (PDA). A unit conversion table will also be necessary.

Section A

Solve for x in terms of y.
1) $y = 5x + 6$
2) $y = \frac{1}{5}x - 3$

Simplify.
3) $y^3 + y^3$
4) $(y^3)(y^3)$
5) $6w^5 - 2w^5$
6) $5x^3 + 2x^5$
7) 6^{-2}
8) $5x^3y^2 - x^3y^2$
9) $(5x^3y^2)(-x^3y^2)$
10) $(5x^3y^2)^2$
11) $(5x^3 - y^2)^2$

Multiply.
12) $5x^2(x + 3)$
13) $(x + 5)(x + 3)$
14) $(x - 5)(x - 1)$
15) $(x + 6)(x - 6)$
16) $(2x + 5)(x - 3)$

Factor.
17) $x^2 + 10x + 16$
18) $x^2 + 5x + 6$
19) $x^2 + 5x - 6$
20) $x^2 - 5x + 6$
21) $x^2 - 5x - 6$
22) $x^2 - 25$
23) $x^2 + 25$
24) $x^2 - 100$
25) $x^2 - 10$
26) $x^2 + 10x + 25$
27) $x^2 - 11x + 18$
28) $x^2 - 49$
29) $x^{12} - 81$
30) $3x^7 + 12x^3$
31) $x^3 - 10x^2 - 24x$

Solve.
32) $5x - 6 = 2x + 21$
33) $8 - 3(x + 7) = x + 7$
34) $x^2 + 6x + 8 = 0$
35) $x^2 + 6x = 40$
36) $13x = x^2 + 30$
37) $\frac{8}{x+2} = \frac{2}{x-4}$
38) $6 - 5(x - 4) = 3(2x + 5) - x$
39) $4x^2 + 3x = 5x^2 + 3x - 1$

Percent Review

40) What is 7% of 5000?

41) 9 is what percent of 43?

42) 23 is what percent of 50?

43) 38 is what percent of 810?

44) What is 450 increased by 80%?

45) What is 180% of 450?

46) What percentage increase is it going from 25 up to 31?

47) What percentage decrease is it going from 31 down to 25?

48) What is 72 decreased by 60%?

49) What is 7000 increased by 3%?

PDA Review

50) Unit Conversions
    (Round to 3 significant digits)

   a) 24 yd = _____ ft

   b) 26 m = _____ cm

   c) 15 kg ≈ _____ lb

   d) 921 ft ≈ _____ m

   e) 2'9" ≈ _____ mm

Section B

Solve for x in terms of y.

51) $7x + 3y = 5$

52) $\frac{2}{3}x - 2y = \frac{3}{4}$

Percent Review

53) 75 is 35% more than what?

54) A bike is on sale for $450. What is the regular price if that sale price is a 20% discount?

PDA Review

55) Unit Conversions

   a) 2500 yd ≈ _____ km

   b) 2½ cups ≈ _____ mℓ

   c) 15 g ≈ _____ oz

   d) 900 cm ≈ _____ ft

   e) 74mm ≈ _____ in

56) A farmer figures that planting a 90-hectare field will produce 225 m³ of wheat. Calculate his yield both in m³/hectare and ft³/acre.

— Midyear Review —
Problem Set #2

Section A
Simplify.
1) $7x^3y^4 + 2x^3y^4$
2) $(4x^{-5})^{-2}$
3) $\dfrac{9x^3z^7}{15x^8y^5z^4}$
4) $5x^6 + 2x^6$
5) $(5x^6)(2x^6)$
6) $6x^3 + 2x^2$
7) $(10x^4)^3$

Multiply.
8) $(x+8)(x-10)$
9) $(2x+1)(3x+4)$
10) $3x^4(2x^5 - 7x)$
11) $(x+4)^2$
12) $(x+9)(x-9)$
13) $(x-9)(x-9)$
14) $(3x-4)(x^2-3x+7)$

Factor.
15) $x^2 - 15x + 44$
16) $x^2 + 10x + 24$
17) $x^2 - 10x + 24$
18) $x^2 + 10x - 24$
19) $x^2 - 10x - 24$
20) $x^6 - 16$
21) $x^5 - 16$
22) $x^2 - 7$
23) $20y^5 + 30y^3$

Solve.
24) $9x + 50 = 4x$
25) $8 - 2(x+7) = 6x - 3$
26) $x^2 - 13x + 40 = 0$
27) $3x = x^2 - 70$
28) $x^2 + 3x = 28$
29) $(x-7)^2 = 2x^2 + 89$
30) $(x-7)^2 = x^2 + 21$
31) $8x + 9 = 1 + 2(x+4)$
32) $2x + 9 = 1 + 2(x+4)$
33) $2x + 5 = 1 + 2(x+4)$
34) $\dfrac{5x}{3x-1} = \dfrac{2x}{x+4}$

Percent Review
35) What is 18% of 62?
36) What is 1.8% of 62?
37) 18 is what percent of 20?
38) 18 is what percent of 200?
39) 18 is what percent of 2000?
40) What percentage increase is it going from 40 up to 70?
41) What percentage increase is it going from 40 up to 80?
42) What percentage increase is it going from 40 up to 120?

DO NOT COPY   DO NOT COPY

— Midyear Review —

43) What percentage increase is it going from 60 up to 135?

44) 135 is what percent of 60?

45) Jim bought some stock for $3,000 and then sold it for $3,600 one year later. What is the profit as a percentage?

46) Jim bought some stock for $3,600 and then sold it for $3,000 one year later. What is the loss as a percentage?

PDA Review

47) Unit Conversions
   a) $0.87\ell =$ _____ mℓ
   b) 10 lb = _____ oz
   c) 900mg = _____ kg
   d) 0.79mm = _____ cm
   e) 400 fl oz = _____ qt
   f) 0.05km = _____ m
   g) $0.62m\ell =$ _____ ℓ
   h) 49m = _____ mm
   i) 1½ ton = _____ oz
   j) $5.8\ell \approx$ _____ gal
   k) 8000 ft $\approx$ _____ km

Section B

Factor.

48) $10x^7 - 120x^6 + 270x^5$

49) $90x^2y^2 - 250x^2y$

PDA Review

50) Unit Conversions
   a) $12 \frac{yd}{s} \approx$ _____ $\frac{km}{h}$
   b) 26 mph $\approx$ _____ $\frac{m}{s}$
   c) 8 lb $\approx$ _____ g
   d) 850 cm $\approx$ _____ yd
   e) 2.8 yd^3 $\approx$ _____ gal

51) What is the density (in lb/ft^3) of a block that weighs 78 lbs and has a volume of 1.3 cubic feet?

52) What is the weight of a block that has a density of 125 lbs/ft^3 and a volume of 0.72 ft^3?

53) What is the volume of a block that weighs 90 lbs and has a density of 230 lbs/ft^3?

— Midyear Review —

Problem Set #3

Section A
Simplify.
1) $12x^3 + 3x^3$
2) $7x^2 + x^2$
3) $7x^2y^4 - x^2y^4$
4) $7x^2y^4 + 3x^3y^4$
5) $(5x^3)(4x^2)$
6) $(5x^3)^2$
7) $\sqrt{900x^6}$

Multiply.
8) $9x^2(x^5 + 3x)$
9) $(x + 2)(x - 1)$
10) $(x - 12)(x - 4)$
11) $(x + 10)(x - 10)$
12) $(y - 6)^2$
13) $(x^4 - 5y^3)(x^4 + 5y^3)$
14) $(x^4 - 5y^3)^2$

Factor.
15) $x^2 + 4x - 21$
16) $x^2 + 13x + 30$
17) $x^2 + 13x - 30$
18) $x^2 - 13x - 30$
19) $x^2 - 13x + 30$
20) $x^2 - 1$
21) $x^{10} - 49$
22) $x^4 + 49$
23) $18x^3y + 24x^2y^5$
24) $x^8 - 10000$
25) $8x^8y^4w^5 - 32x^2y^4w^5$

Solve.
26) $4x - 1 = 10x + 23$
27) $13x = 41 - (-2x - 3)$
28) $0 = x^2 + 5x - 14$
29) $x^2 + x = 56$
30) $5x^2 + 3x - 11 = 6x^2 + 15x + 9$

Percent Review
31) What is 38% of 247?
32) What is 0.4% of 3000?
33) What is 708 increased by 13.8%?
34) 5.3 is what percent of 660?
35) What do you end up at when 8000 is increased by 25% and then that result is decreased by 25%?
36) What is 200 decreased by 30%?
37) A bike normally listed for $450 is on sale for a 20% discount. What is the new discounted price?

PDA Review
38) Unit Conversions
 a) 5.8 m = _____ cm
 b) 81 mℓ = _____ ℓ
 c) 5 qt = _____ gal
 d) 3 mi = _____ ft
 e) 8 ft ≈ _____ cm
 f) 5½ lb ≈ _____ kg
 g) 0.39 ℓ ≈ _____ fl oz

— Midyear Review —

Section B

Simplify.

39) $(y^2 - 3)^3$

Factor.

40) $x^2 + 34x + 240$

41) $x^2 + 34x - 240$

42) $x^2 - 34x + 240$

43) $x^2 - 34x - 240$

Solve for x in terms of y.

44) $y = -2x + \frac{3}{4}$

45) $\frac{1}{2}x - \frac{2}{3}y = 7$

Solve.

46) $(x - 5)^2 = x(x - 10)$

47) $4x^2 - 6(x-1) = 7x^2 - 39$

Percent Review

48) 24¾ is 45% of what?

49) 527 is 15% less than what?

50) Hank is 60% as tall as Betty.
   a) How tall is Hank if Betty is 120cm tall?
   b) How tall is Betty if Hank is 120cm tall?

51) Betty is 60% taller than Hank.
   a) How tall is Hank if Betty is 120cm tall?
   b) How tall is Betty if Hank is 120cm tall?

PDA Review

52) Unit Conversions
   a) 0.7 oz ≈ _____ mg
   b) 200 km² ≈ _____ mi²
   c) 200 mi² ≈ _____ km²
   d) 6.3 $\frac{mi}{min}$ ≈ _____ $\frac{m}{s}$
   e) 9000 mm ≈ _____ ft
   f) 344 $\frac{lb}{ft^3}$ ≈ _____ $\frac{kg}{m^3}$

53) Which is faster, 28 mph or 38 ft/s?

54) How much does a cube of aluminum weigh that has 8-inch long edges?

55) What is the volume of 300 grams of mercury? Give your answer both in milliliters and in fluid ounces.

56) What is the density (in g/cm³) of a cube that weighs 7.3 kg and has edges that are 6 cm long?

— Midyear Review —

Problem Set #4

Section A

Simplify.
1) $8z^4 - 5z^4$
2) $8z^4 + 5z^8$
3) $(8z^4)(5z^8)$
4) $3x^2y^6 - 13x^2y^6$
5) $(3x^2y^6)(-13x^2y^6)$
6) $(3x^4y^3)^2$
7) $\dfrac{5x^{-3}y^{-4}}{3x^{-5}y^2}$
8) $(\frac{3}{4})^{-3}$
9) $\sqrt{2500x^6y^4}$

Multiply.
10) $(x+8)(x+4)$
11) $(x-10)(4x-3)$
12) $(x^5+2)(x^5-2)$
13) $(x+5y)(x-7y)$
14) $(x^4+3)^2$
15) $6y^3(3y^2-7y)$

Factor.
16) $x^2 - 2x - 35$
17) $x^2 + 17x + 60$
18) $x^2 + 17x - 60$
19) $x^2 - 17x + 60$
20) $x^2 - 17x - 60$
21) $x^2 - 4$
22) $x^4 - 81$
23) $3x^5 + 6x^4 - 24x^3$
24) $7x^3 - 28x$
25) $8x^9y^4 - 18x^3y^4$

Solve.
26) $7x - 3 = x + 27$
27) $0 = x^2 - x - 42$
28) $x^2 + 24x = 2x - 40$
29) $4 - 3(x+6) = -17x + 12x$
30) $4 - 3(x+6) = x^2 + 12x$
31) $(x+8)(x-3) = x^2 - 24$
32) $\frac{4}{5}x - \frac{1}{2} = \frac{2}{5}x + \frac{3}{5}$

Percent Review
33) 170 is what % of 6000?
34) What is 120% of 45?
35) What is 45 increased by 20%?
36) What percentage increase is it going from 1700 up to 2100?
37) What percentage decrease is it going from 400 down to 100?
38) What is 4600 decreased by 90%?
39) What is 10% of 4600?
40) In a local election for mayor, with approximately 38,000 people voting, 62% of the votes were cast in favor of Joe. Approximately how many people didn't vote for Joe?

— Midyear Review —

PDA Review
41) Unit Conversions
   a) 24 fl oz = _____ cups
   b) 3.9 m ≈ _____ ft
   c) 7.4 ℓ = _____ mℓ
   d) 1400 mm = _____ km
   e) 570 m = _____ km
   f) 3¼ cups ≈ _____ mℓ

Section B
Factor.
42) $4y^5 - 40y^4x^2 + 36y^3x^4$
43) $8x^2 - 83x + 30$
44) $8x^2 + 56x - 30$
45) $8x^2 - x - 30$

Solve.
46) $2x^2(2x-5)^2 + 72 = 20x^2(5-2x)$

Percent Review
47) 2600 is 30% more than what?
48) 2600 is 130% of what?
49) Keith has 80% more money than Greta. How much money does Keith have if Greta has $990?
50) Keith has 80% more money than Greta. How much money does Greta have if Keith has $990?
51) Jeff left a $8.03 tip, which was 22% of the meal's price. What was the price of the meal?
52) A jacket is marked at a discounted price of $33. If this was a 45% discount, what was the original price?

PDA Review
53) Unit Conversions
   a) 17.3 ℓ ≈ _____ pt
   b) 4 ft³ = _____ in³
   c) 4 ft³ = _____ yd³
   d) $70 \frac{kg}{m^3}$ = _____ $\frac{g}{cm^3}$
   e) $13 \frac{ft}{sec}$ ≈ _____ mph

54) A car has a fuel efficiency of 25 km/ℓ. What is this in mpg?

55) A concrete block measures 30cm by 25cm by 20cm. What does the block weigh (in kg) if the density of concrete is 2.1 g/cm³?

56) What is the volume (in both in³ and ft³) of 100 pounds of gold?

57) A 300g block of cheese in England costs £1.74. In Germany a 250g block of cheese costs 1.45 euros. In the U.S. a 9-ounce block of cheese costs $1.79. The exchange rates are as follows:
$1 = £0.578 = 0.855euro
   a) Cheese is what percent more expensive in England than the U.S.?
   b) Cheese is what percent cheaper in Germany than the U.S.?

58) If the ratio of boys to girls is 3:2, then how many of each are there if there are 150 children?

— Midyear Review —
Problem Set #5

Section A
Simplify.
1) $5x^7 - x^7$
2) $4x^7 + 5x^6$
3) $(4x^7)(5x^6)$
4) $(2x^5)^3$
5) $\dfrac{4x^{-3}y^{-2}}{7y^5z^{-4}}$
6) $\left(\dfrac{5x^{-2}}{4y^3}\right)^{-3}$
7) $6x^5y^2 - y^2$
8) $(5x^2y)(3x^4y^3)^2$
9) $(x^6+4)(x^3+2)(x^3-2)$

Multiply.
10) $(x+6)(x+9)$
11) $(x^3 - 11)(x^3 - 9)$
12) $(x^2 + 8)(x^2 - 8)$
13) $(2x - 5)(3x - 1)$
14) $-4x^5(2x^4 - 3x^2)$
15) $(w + 9)^2$
16) $6x(x + 5)^2$

Factor.
17) $x^2 + 7x + 6$
18) $x^2 - 25x + 150$
19) $x^2 - 25x - 150$
20) $x^2 + 25x - 150$
21) $x^2 + 25x + 150$
22) $x^2 - 144$
23) $x^2 + 144$
24) $x^3 - 9$
25) $10x^3y^7 + 8x^2y^4$
26) $5x^7 - 45x^5$
27) $2x^3 + 14x^2 + 24x$
28) $7x^6 - 21x^3$
29) $x^2 - 17x + 70$
30) $3x^5 - 12x^3$

Solve.
31) $5x - 6 - x = 9 - 10x - 22$
32) $5 - 4(2x + 3) = 6 - (4x + 5)$
33) $x^2 + 16x + 48 = 0$
34) $5x^3 + 20x^2 - 25x = 0$
35) $x^2 + 37x = 16 + 37x$
36) $3x^2 + 10x = 2x^2 - 25$

Percent Review
37) What is 0.03% of 3400?
38) 6 is what percent of 8?
39) What is 350% of 4000?
40) What is 4000 increased by 250%?
41) What percentage increase is it going from 210 up to 575?
42) 73 is 27% of what?

— Midyear Review —

PDA Review

43) Unit Conversions
   a) 30 ft = _____ yd
   b) 90 kg = _____ mg
   c) 9 cm = _____ mm
   d) 7 lb ≈ _____ kg
   e) 700 cm ≈ _____ in
   f) 3000 oz ≈ _____ kg

44) If a model of the Earth were made exactly to scale with a diameter of one meter, how far above the surface of the model would Mount Everest stick out? (Mount Everest has a height of about 8800m and the Earth has a radius of about 6400km.)

Section B

Factor.

45) $x^3 - x^5$
46) $5x^8 - 30x^7 + 40x^6$
47) $4x^6 - 9y^8$

Solve.

48) $3x^2 + 5 = (x+7)^2 + 16$
49) $3x^3(x+3)^2 = 6x^3(3x+17)$
50) $x^4 - 9x^2 = 4(x-3)(x+3)$

Percent Review

51) 121 is 12% less than what?

52) George has 80% as much money as Vicky. How much money does George have if Vicky has $990?

53) George has 80% as much money as Vicky. How much money does Vicky have if George has $990?

54) Bob weighs 20% more than Pete.
   a) Bob's weight is what percent of Pete's?
   b) Pete's weight is what percent of Bob's?
   c) Pete weighs what percent less than Bob?

PDA Review

55) Unit Conversions
   a) $14 \text{ ft}^3 \approx$ _____ m^3
   b) $26 \tfrac{m}{s} =$ _____ $\tfrac{km}{h}$
   c) $800 \tfrac{in}{s} \approx$ _____ mph

56) What is the volume (in in^3) of a block of iron that weighs 10 pounds?

57) A rock has a volume of 5.6 ft^3 and weighs 1400 pounds.
   a) What is the density in both lb/ft^3 and oz/in^3?
   b) What percent as dense as gold is it?
   c) What percent as dense as water is it?

58) Hans bought a 3.7-hectare plot of land in Germany for 3.2 million euros. What is the cost of this land in dollars per acre? ($1 = 0.855$ euro)

— Midyear Review —

Problem Set #6

Section A
Simplify.
1) $7x^6 - x^6$
2) $3y^2 + 5y^2$
3) $3y^2 + 5x^2$
4) $(3y^2)(5x^2)$
5) $(3y^2)^4$
6) $\left(\frac{2x^2 y^{-3}}{3y^3 z^{-4}}\right)^{-2}$
7) $7x^5 y^2 + x^5 y^2$
8) $3(x+2y)(5x-3y)$

Multiply.
9) $(x+7)(x+2)$
10) $(x^3 + 7y)(x^3 + 2y)$
11) $(w^4 + 5)(w^4 - 5)$
12) $(6x-5)(2x+3)$
13) $2y^2(y+3)(y-6)$
14) $(x^3 - 4)^2$

Factor.
15) $x^2 - 12x + 11$
16) $x^2 - x - 90$
17) $x^2 - 34x - 240$
18) $x^2 + 34x + 240$
19) $x^2 + 34x - 240$
20) $x^2 - 34x + 240$
21) $10x^5 - 90x^4 + 180x^3$
22) $x^8 - 9$
23) $x^8 + 9$
24) $x^8 + 9x^6$
25) $x^7 - 9x$
26) $x^8 - 1$

Solve.
27) $6 - x = 8x + 7$
28) $x^2 - 5x = 24$
29) $x^2 - 4x - 21 = 2x^2 - 18$
30) $(x+1)(x+4) = x^2$
31) $(x+1)(x+4) = 40$
32) $8 - 3x = x^2 + 4$
33) $8 - 3x = x + 4$
34) $\frac{3x+6}{12} = \frac{x+2}{4}$

Percent Review
35) What is 93.2% of 8000?
36) 7 is what percent of 23?
37) 10 is what percent of 15?
38) 15 is what percent of 10?
39) What percentage increase is it going from 73 up to 90?

PDA Review
40) Unit Conversions
   a) 68 mg = _____ kg
   b) 24 g ≈ _____ oz
   c) 4 gal = _____ pt
   d) 3140 km ≈ _____ mi
   e) 200 mℓ ≈ _____ fl oz
   f) 2.1 km ≈ _____ yd

Section B

Solve.

41) $3(x+2)(x+5) = 3x(x-3)$

42) $(x-4)^3 = 2(x^2-32)$

43) $2x^5(x+4)(x-2) = 4x^6 + 18x^3$

Percent Review

44) 75 is 20% of what?

45) 75 is 20% more than what?

46) 75 is 20% less than what?

47) In an election with 3,600 people voting and only two candidates running, the loser received 40% of the votes. How many votes did each candidate receive?

48) In an election with 3,600 people voting and only two candidates running, the loser received 40% fewer votes than his opponent. How many votes did each candidate receive?

49) In an election with 3,600 people voting and only two candidates running, the winner received 40% more votes than his opponent. How many votes did each candidate receive?

50) Sally has 80% less money than Mark. How much money does Sally have if Mark has $990?

51) Sally has 80% less money than Mark. How much money does Mark have if Sally has $990?

PDA Review

52) Unit Conversions
   a) $0.9\, \ell \approx$ _____ cups
   b) $130\, \frac{km}{h} \approx$ _____ mph
   c) $4{,}000{,}000\text{ cm}^3 \approx$ _____ ft^3
   d) $2.5\, \frac{yd}{sec} \approx$ _____ $\frac{km}{h}$
   e) $43\text{ ft}^3 \approx$ _____ ℓ
   f) $14.7\, \frac{lb}{in^2} \approx$ _____ $\frac{kg}{m^2}$

53) A school has an enrollment of 495, and the ratio of boys to girls is 7:8. How many boys are there in the school?

54) In 2005, Bill Carpenter set a record for running the Leadville 100 (a 100-mile race) in just 15 hours and 42 minutes.
   a) What was his average speed in miles per hour?
   b) What was his average speed in miles per minute?
   c) How many minutes per mile is this?

55) A 4-pint container of milk in England costs £1.05. In Germany a liter of milk costs 0.53 euros. In the U.S. a gallon of milk costs $2.50. The exchange rates are: $1 = £0.578 = 0.855$ euro
   a) Milk in England is what percent more expensive than milk in the U.S.?
   b) Milk in Germany is what percent cheaper than milk in the U.S.?

56) A block has a volume of 59 in^3, and weighs 7.8 lb. Find the density of the block both in lb/ft^3 and kg/m^3.

Fractions & Square Roots

Problem Set #1

Group Work
Simplifying Square Roots

One of the goals when simplifying a square is to "remove" any perfect squares from inside the square root. The following example will demonstrate how this is done:

Example: Simplify $\sqrt{45}$

Solution: We see that 9, which is a perfect square, is a factor of 45. Therefore:

$$\sqrt{45} \to \sqrt{9 \cdot 5} \to \sqrt{9} \cdot \sqrt{5}$$
$$\to 3\sqrt{5}$$

Note: A square root is simplified once the number inside the square root sign has no factors that are perfect squares. The perfect squares are 4, 9, 16, 25, etc.

Simplify each square root.

1) $\sqrt{50}$
2) $\sqrt{700}$
3) $\sqrt{72}$
4) $\sqrt{1296}$

Simplifying Rational Expressions (fractions).

When simplifying rational expressions, you cannot cancel if multiple terms appear in the numerator or denominator. With the example below, you cannot simply cancel $6x^3$ from the top and bottom. There are two ways to do this problem.

Example: $\dfrac{36x^7 + 24x^5 + 6x^3}{6x^3}$

Solution #1:
$\dfrac{36x^7}{6x^3} + \dfrac{24x^5}{6x^3} + \dfrac{6x^3}{6x^3} \to 6x^4 + 4x^2 + 1$

Solution #2:
$\dfrac{6x^3(6x^4 + 4x^2 + 1)}{6x^3} \to 6x^4 + 4x^2 + 1$

5) $\dfrac{9x^4 - 6x^3}{3x}$

6) $\dfrac{16x^4y^2 - 12x^2y^4}{4x^2y^2}$

7) $\dfrac{4x^5 + 12x^4 - 20x^3}{4x^3}$

Homework
Simplify.

8) $\sqrt{175}$
9) $\sqrt{12}$
10) $\sqrt{120}$
11) $\sqrt{1200}$
12) $\sqrt{12000}$
13) $\sqrt{120000}$
14) $\sqrt{32}$
15) $\sqrt{324}$
16) $\sqrt{3240}$
17) $\sqrt{32400}$

18) $\dfrac{21x^3 + 7x^2}{7x^2}$

19) $\dfrac{15x^3y^2 + 25x^4y^3}{5xy^2}$

20) $\dfrac{6x^6 - 9x^4 + 3x^2}{3x^2}$

— Fractions & Square Roots —

Problem Set #2

Group Work
Simplify.
1) $\sqrt{8}$
2) $\sqrt{80}$
3) $\sqrt{800}$
4) $\sqrt{8000}$
5) Use a calculator to give a decimal approximation:
   a) $\sqrt{50}$
   b) $5\sqrt{2}$
   c) $\sqrt{300}$
   d) $10\sqrt{3}$
   e) What do the above four answers show us?
6) $\sqrt{2^4 \cdot 3^2 \cdot 5}$
7) $\sqrt{2^4 \cdot 3^2 \cdot 5^3}$
8) $\sqrt{2^5 \cdot 3^4 \cdot 5^3}$
9) $\dfrac{3x^4 - 6x^3 - 18x^2}{3x}$
10) $\dfrac{x^2 + 8x + 12}{x + 6}$
11) $\dfrac{x^2 - x - 12}{x^2 - 9}$
12) $\dfrac{2}{2 - \frac{2}{3}}$
13) $\dfrac{2}{2 - \frac{2}{x}}$

Homework
Simplify.
14) $\sqrt{490}$
15) $\sqrt{44}$
16) $\sqrt{45}$
17) $\sqrt{450}$
18) $\sqrt{3^5}$
19) $\sqrt{3^6}$
20) $\sqrt{3^2 \cdot 5^4 \cdot 11^2}$
21) $\sqrt{3^2 \cdot 5^3 \cdot 11^3}$
22) $\dfrac{12x^3y^2 - 10x^2y^4}{2xy}$
23) $\dfrac{2x^3 - 8x^2 - 24x}{2x}$
24) $\dfrac{2x^3 - 8x^2 - 24x}{x + 2}$
25) $\dfrac{x^2 + 2x - 24}{x^2 - 6x + 8}$
26) $\dfrac{x^2 - 4x - 5}{x^2 - 25}$
27) $\dfrac{x^3 - 6x^2 + 9x}{4x^2 + 8x - 60}$
28) $\dfrac{18x^5 - 15x^3}{3x^2}$
29) $\dfrac{x^2 - 16}{x^2 + x - 20}$
30) $\dfrac{3}{3 - \dfrac{3}{3 - \frac{3}{x}}}$
31) $\dfrac{1 + \frac{1}{x}}{1 - \frac{1}{x^2}}$
32) The difference of two numbers is four and the sum of their squares is 58. What are the two numbers?

— Fractions & Square Roots —

Problem Set #3

Group Work

1) One rule for simplifying square roots is that a square root is not allowed in the denominator. How then can you simplify $\frac{3}{\sqrt{5}}$?

Simplify.

2) $\frac{5}{\sqrt{2}}$

3) $\frac{6}{\sqrt{7}}$

4) $\frac{\sqrt{3}}{\sqrt{5}}$

5) $\frac{\sqrt{6}}{\sqrt{3}}$

6) $\frac{x+4}{4+x}$

7) $\frac{x-4}{4-x}$

8) $\frac{x+4}{4-x}$

9) $\frac{2}{5x^2} - \frac{3}{10x}$

10) $\frac{3}{x+4} + \frac{4}{x+1}$

11) $\frac{5x}{2x+6} - \frac{3}{x^2+3x}$

Find the reciprocal.
(Give simplified answers.)

12) $\frac{x-3}{x^2}$

13) $\sqrt{5}$

14) $\frac{\sqrt{3}}{3}$

Homework
Simplify.

15) $\sqrt{48}$

16) $\sqrt{30}$

17) $\sqrt{2520}$

18) $\frac{5}{\sqrt{2}}$

19) $\frac{6}{\sqrt{6}}$

20) $\frac{\sqrt{5}}{\sqrt{7}}$

21) $\frac{\sqrt{15}}{\sqrt{5}}$

22) $\frac{3\sqrt{2}}{2\sqrt{10}}$

23) $\frac{20x^3y^4 - 15x^4y^7 + 10x^2y^2}{5xy^2}$

24) $\frac{14x}{21x^4}$

25) $\frac{x^2 - 10x + 21}{x^2 + 10x - 39}$

26) $\frac{2x^3 - 16x^2 + 24x}{x^2 + 4x - 12}$

27) $\frac{x^2 - 9}{x^2 - 4x + 3}$

28) $\frac{9 - x^2}{x^2 - 4x + 3}$

29) $\frac{1 - \frac{4}{x^2}}{x + 2}$

30) $\frac{c + \frac{3c}{c-3}}{c - \frac{3c}{c+3}}$

31) $\frac{5}{2x^2y^3} + \frac{3}{4xy^5}$

32) $\frac{2}{x-2} + \frac{3}{x^2 - 4}$

33) $\frac{6}{x+5} + \frac{2}{x-3}$

34) Find the common solution
$4x - 3y = 18$
$2x + 5y = -17$

— Fractions & Square Roots —

Problem Set #4

Group Work
Solve.
1) $\dfrac{8}{4x-3} = \dfrac{1}{x} + \dfrac{1}{x-2}$

Homework
Give the Reciprocal.
2) $\sqrt{13}$
3) $\dfrac{2}{\sqrt{2}}$

Simplify.
4) $7\sqrt{3} + 8\sqrt{3}$
5) $5\sqrt{6} - 8\sqrt{6}$
6) $5\sqrt{6} + 3\sqrt{7}$
7) $\sqrt{12} + \sqrt{27}$
8) $(2\sqrt{5})^2$
9) $(3\sqrt{6})^2$
10) $\dfrac{x+7}{7+x}$
11) $\dfrac{x-3}{x+3}$
12) $\dfrac{x-9}{9-x}$
13) $\dfrac{3x^4-5}{5-3x^4}$
14) $\dfrac{6x^5}{8x}$
15) $\dfrac{7}{x+2} + \dfrac{3}{x-5}$
16) $\dfrac{7}{x+2} - \dfrac{3}{x-5}$
17) $\dfrac{5}{x+5} + \dfrac{4}{x-5}$
18) $\dfrac{5}{5-x} + \dfrac{4}{x-5}$
19) $\dfrac{\dfrac{1}{m} - \dfrac{1}{2m}}{\dfrac{2}{m}}$
20) $\dfrac{\dfrac{8x^3-8x}{x^2-2x-3}}{\dfrac{10x^6-10x^5}{x^2+x-12}}$
21) $\dfrac{x - \dfrac{1}{2x+1}}{1 - \dfrac{2}{2x+1}}$
22) $\dfrac{\dfrac{x}{y} - \dfrac{y}{x}}{\dfrac{1}{2x} - \dfrac{1}{2y}}$

Solve.
23) $\dfrac{2}{x} + \dfrac{3}{x-1} = 4$
24) $\dfrac{3}{2x-1} - \dfrac{2x+1}{3} = 2$
25) $\dfrac{3}{x+4} = \dfrac{2}{x^2-16} + \dfrac{1}{x-4}$
26) $\dfrac{1}{2x-6} - \dfrac{1}{3x-6} = \dfrac{x-1}{x^2-5x+6}$
27) The sum of two numbers is 5. One of the numbers squared is 7 more than the other. What are the two numbers?

— Fractions & Square Roots —

Problem Set #5

Group Work
Simplify.
1) $(2 + \sqrt{3})(7 + \sqrt{3})$
2) $(4 + 3\sqrt{2})(5 - \sqrt{2})$
3) $(3 + \sqrt{5})^2$
4) $(5 - 4\sqrt{3})^2$
5) $(3 - \sqrt{2})(3 + \sqrt{2})$

Homework
Simplify.
6) $\sqrt{50} + \sqrt{18}$
7) $(6 + \sqrt{3})(2 + \sqrt{3})$
8) $(6 + \sqrt{2})(2 + \sqrt{3})$
9) $(2 + \sqrt{7})^2$
10) $(3 - 2\sqrt{5})^2$
11) $(5 - \sqrt{3})(5 + \sqrt{3})$
12) $\dfrac{4}{3\sqrt{5}}$
13) $\dfrac{5}{3 + \sqrt{2}}$
14) $\dfrac{3}{4x^2y} - \dfrac{1}{3xy^3}$
15) $\dfrac{4}{x^3} + \dfrac{2}{x^2y^2} - \dfrac{1}{y^2}$
16) $\dfrac{4}{x - 5} + \dfrac{1}{5 - x}$
17) $\dfrac{2x}{x - 3} - \dfrac{3}{x + 3}$
18) $xy^{-1} + x^{-1}y$
19) $\dfrac{6x - 5}{5 - 6x}$
20) $\dfrac{4x^3 - 20x^2}{25 - x^2}$
21) $\dfrac{3x^2 + 6x}{4 - x^2}$
22) $\dfrac{2}{2 - \dfrac{2}{2 - \frac{1}{2}}}$
23) $\dfrac{\dfrac{x}{x+y} + \dfrac{y}{x-y}}{\dfrac{x}{x-y} - \dfrac{y}{x+y}}$

Solve.
24) $\dfrac{1}{x} + \dfrac{1}{x+5} = \dfrac{1}{6}$
25) $2 - \dfrac{5}{x^2 - x - 6} = \dfrac{x+3}{x+2}$
26) The difference of two numbers is $5/12$ and their product is 6. Find the two numbers.

— Fractions & Square Roots —

Problem Set #6

Group Work
Simplify.

1) $\dfrac{\sqrt{21}}{\sqrt{15}}$

2) $\dfrac{6}{5\sqrt{2}}$

3) $\dfrac{6}{5+\sqrt{2}}$

4) $\dfrac{3-\sqrt{2}}{5+\sqrt{2}}$

5) Let $x = 4 - \sqrt{3}$ and $y = 4 + \sqrt{3}$. x and y are called *conjugates*. Find:
   a) x and y on a calculator.
   b) x+y
   c) x·y
   d) What is special about conjugates?

6) Find the common solution to
   $3x - y = 12$
   $2y - x^2 + 6x = 11$

Homework
Simplify.

7) $\sqrt{150}$

8) $\dfrac{6}{5\sqrt{3}}$

9) $\dfrac{3\sqrt{6}}{4\sqrt{15}}$

10) $(4+\sqrt{2})^2$

11) $(4\sqrt{2})^2$

12) $(4+\sqrt{2})(4-\sqrt{2})$

13) $(1 - 2\sqrt{3})^2$

14) $\dfrac{3}{2-\sqrt{3}}$

15) $\dfrac{6}{5-\sqrt{7}}$

16) $\dfrac{6-\sqrt{6}}{6-\sqrt{2}}$

17) $\dfrac{x}{x+3} - \dfrac{4}{x-3}$

18) $\dfrac{x}{3-x} + \dfrac{3}{x-3}$

19) $\dfrac{3}{4x+6} - \dfrac{x+4}{6x+9}$

20) $\dfrac{8x^3y^4 - 6x^4y^2 + 2x^2y^2}{2x^2y^2}$

21) $\dfrac{\frac{3x^3}{x^2-4}}{\frac{6x^2-9x}{2x^2+x-6}}$

22) $\dfrac{1}{1 - \dfrac{1}{1 - \dfrac{1}{1-\frac{1}{x}}}}$

Solve.

23) $\dfrac{6x}{x+15} = \dfrac{1}{4-x}$

24) $\dfrac{x+1}{2x-2} = \dfrac{x}{6} + \dfrac{1}{x-1}$

25) $\dfrac{3x}{x-1} - \dfrac{4}{x+1} = \dfrac{4}{x^2-1}$

— Fractions & Square Roots —

Problem Set #7

Group Work
Long Division
Study the example below. It shows that $(x^3+5x^2+11x+10) \div (x+2)$ equals x^2+3x+5.

$$\begin{array}{r} x^2 + 3x + 5 \\ x+2 \overline{\smash{)}x^3+5x^2+11x+10} \\ -\underline{(x^3+2x^2)} \\ 3x^2+11x \\ -\underline{(3x^2+6x)} \\ 5x+10 \\ -\underline{(5x+10)} \\ 0 \end{array}$$

Now try this one:

1) $\dfrac{x^3+6x^2+15x+28}{x+4}$

Homework
Simplify.

2) $\sqrt{280}$

3) $\sqrt{3^5}$

4) $3\sqrt{2} + 4\sqrt{5}$

5) $(3\sqrt{2})(4\sqrt{5})$

6) $\dfrac{9}{5\sqrt{7}}$

7) $\dfrac{2x^2}{8x^6}$

8) $\sqrt{3} \cdot \dfrac{\sqrt{3}}{3}$

9) $(2-\sqrt{5})^2$

10) $(3+2\sqrt{5})^2$

11) $(4-2\sqrt{3})(4+2\sqrt{3})$

12) $\dfrac{9}{5+\sqrt{7}}$

13) $\dfrac{2+\sqrt{3}}{2-\sqrt{3}}$

14) $\dfrac{3x}{x+7} + \dfrac{3}{x-4}$

15) $\dfrac{7}{x+3} - \dfrac{4}{3+x}$

16) $\dfrac{7}{x-3} - \dfrac{4}{3-x}$

17) $\dfrac{7}{x-3} - \dfrac{4}{x+3}$

18) $\dfrac{x-2}{x^2-25} - \dfrac{2}{3x^2+15x}$

19) $\dfrac{\frac{x}{y} - \frac{x-y}{x+y}}{\frac{y}{x} + \frac{x-y}{x+y}}$

20) $\dfrac{\frac{4x^2-16}{15x^2-30x+15}}{\frac{12-6x}{25x^2-25}}$

Divide.

21) $\dfrac{x^3+9x^2+23x+15}{x+3}$

Solve.

22) $\dfrac{3x+5}{6} - \dfrac{5}{x} = \dfrac{x}{2}$

23) $\dfrac{2}{x-1} = \dfrac{3}{x-2} + \dfrac{2}{x-4}$

24) $\dfrac{3x-1}{x} + \dfrac{3}{x-3} = \dfrac{9}{x^2-3x}$

25) Janet earns $22/hr as a computer programmer and $16 per hour as a lab assistant. Last week, she worked twice as many hours in the lab as she did programming. How much time did she work in the lab if she made a total of $648 between the two jobs?

— Fractions & Square Roots —

Problem Set #8

Group Work
Divide.

1) $\dfrac{x^3-x^2-11x+15}{x-3}$

2) $\dfrac{6x^3+x^2-10x+3}{2x+3}$

Homework
Simplify.

3) $(5\sqrt{10})^2$

4) $\dfrac{3\sqrt{2}}{4\sqrt{3}}$

5) $(1+\sqrt{5})^2$

6) $(2\sqrt{3}-3\sqrt{5})^2$

7) $\left(\dfrac{2\sqrt{5}}{3}\right)^{-2}$

8) $\dfrac{3}{2+\sqrt{6}}$

9) $\dfrac{3+\sqrt{7}}{2-\sqrt{7}}$

10) $\dfrac{x^2+5x-6}{1-x^2}$

11) $\dfrac{x^2-5x-6}{1-x^2}$

12) $\dfrac{3x}{5y^4}+\dfrac{4}{3x^2y^2}$

13) $4y^{-2}-3xy^{-3}$

14) $x^2 - \dfrac{x-\frac{9}{x}}{\frac{1}{3x^2}+\frac{1}{x^3}}$

15) $\dfrac{x}{x-\frac{x}{1-\frac{1}{x}}}$

Divide.

16) $\dfrac{x^3+11x^2+32x+12}{x+6}$

17) $\dfrac{6x^3-5x^2-19x+20}{3x-4}$

18) $\dfrac{6x^3-5x^2-19x+20}{2x^2+x-5}$

Solve.

19) $\dfrac{1}{x-3}+\dfrac{1}{x+5}=\dfrac{x+1}{x-3}$

20) $\frac{1}{2}x^2-\frac{1}{3}x=2x+4$

21) $\dfrac{5x^2+6}{x^2-4}-\dfrac{3}{x-2}=\dfrac{5x+3}{x+2}$

22) The sum of the squares of three consecutive even integers is 200. Find the numbers.

23) The sum of Tim and Wendy's ages is 30. Three years ago, Tim was three times as old as Wendy. How old is Wendy now?

— Fractions & Square Roots —

Problem Set #9

Group Work
Since $(x+2)(x-5)(x-4)$ multiplies out to become $x^3-7x^2+2x+40$, we can imagine that there would be some way to factor $x^3-7x^2+2x+40$ to become $(x+2)(x-5)(x-4)$. But how? See if you can figure out a way to factor the following problem into three binomials. (Hint: Use long division.)

1) $x^3+7x^2+7x-15$

Homework
Simplify.

2) $\dfrac{8}{3\sqrt{7}}$

3) $\dfrac{8}{3+\sqrt{7}}$

4) $\dfrac{7\sqrt{6}}{\sqrt{3}}$

5) $\dfrac{2-\sqrt{3}}{4-\sqrt{3}}$

6) $(5-2\sqrt{3})^2$

7) $\left(\dfrac{5\sqrt{3}}{3}\right)^{-1}$

8) $\sqrt{45}-\sqrt{500}$

9) $\dfrac{3x}{4y^3}-\dfrac{1}{6x^2y}$

10) $3x^{-2}+5xy^{-3}$

11) $\dfrac{7}{5x+15}-\dfrac{1}{4x^2+12x}$

12) $\dfrac{x^2-6x+11}{6x-x^2-11}$

13) $\dfrac{x^2+6x-11}{x^2+11-6x}$

14) $\dfrac{\frac{72x^3-2x^5}{x^2+x-30}}{\frac{6x^2-36x}{x^2-3x-10}}$

15) $\dfrac{1-\frac{4}{x^2}}{\frac{x}{2}+1}$

Divide.

16) $\dfrac{2x^4-6x^3-19x^2+23x-6}{x^2-5x+2}$

17) $\dfrac{18x^3-17x+6}{3x-2}$

Find the common solution.

18) $6x+4y=1$
    $3x-8y=8$

Factor.

19) $6x^3-49x^2+98x-15$
    given $(x-5)$ is a factor.

20) $10x^3+19x^2-39x-18$
    given $(5x+2)$ is a factor.

Solve.

21) $\dfrac{x}{x+1}=\dfrac{x+1}{x-4}+\dfrac{5}{x^2-3x-4}$

22) $\dfrac{3x}{2x+2}-\dfrac{5}{8}=\dfrac{3x^2}{x^2-1}-\dfrac{23}{4x-4}$

23) The sum of two numbers is 17. The sum of their squares is 91 more than their product. Find the two numbers.

— Fractions & Square Roots —

Problem Set #10

Simplify.

1) $\sqrt{363}$
2) $\sqrt{3^6 \cdot 5^5}$
3) $\sqrt{128} + \sqrt{12} - \sqrt{18}$
4) $\dfrac{7}{\sqrt{2}}$
5) $\dfrac{3 - \sqrt{5}}{\sqrt{5}}$
6) $\dfrac{6 - 2\sqrt{2}}{4 + 3\sqrt{2}}$
7) $\dfrac{15x^2}{5x^5}$
8) $\dfrac{8x^4y^2 - 6x^3y^3}{2xy^2}$
9) $\dfrac{4x^2 + 5x - 6}{4 - x^2}$
10) $\dfrac{3}{4x^2y} + \dfrac{5}{6xy^3}$
11) $\dfrac{5}{x-7} - \dfrac{x+3}{7-x}$
12) $\dfrac{\frac{25}{x} - x}{\frac{x}{2x^2 - 5x}} - 3$
13) $\dfrac{\frac{1}{2x-2} - \frac{1}{x}}{\frac{2}{x} - \frac{1}{x-1}}$

Find the Reciprocal.

14) $\sqrt{11}$
15) $\dfrac{\sqrt{5}}{5}$

Divide.

16) $\dfrac{8x^3 - 24x^2 - 2x + 30}{2x^2 - 3x - 5}$
17) $\dfrac{8x^3 - 27}{2x - 3}$

Give two solutions to:

18) $3x - 4y = 24$

Find the common solution.

19) $7x - 6y = 15$
    $3x - 8y = 20$

Factor.

20) $x^3 + 20x^2 - 4x - 80$
    given $(x+20)$ is a factor.
21) $2x^3 - 14x + 12$
    given $(x-1)$ is a factor.
22) $x^4 + 3x^3 - 12x - 16$
    given $(x^2 - 4)$ is a factor.
23) **Solve.** $\dfrac{x+2}{x-1} - \dfrac{4-x}{2x} = \dfrac{7}{3}$

24) Jeff leaves home at 8:50am on his bike. At 9:10am, Dan leaves the same house on his moped to catch Jeff. At what time does Dan catch up to Jeff if Dan's speed is 50% greater than Jeff's?

25) A pile of 100 coins, worth $8.36, consists of pennies, nickels, dimes and quarters. There are five times as many nickels as pennies, and nine more dimes than pennies. How many of each kind of coin are there?

The Quadratic Formula

Problem Set #1

Group Work

At the end of the *Factoring* unit you were given the following problem to solve:
$$x^2 + 6x = 3$$
However, at that time, you were not able to solve it. (Can you see why?) One of the goals of this new unit is to be able to solve problems like this, and to develop a formula, called the *Quadratic Formula*, that allows us to easily solve these problems. We will now take the first step toward this goal.

Absolute Values

Loosely speaking, taking the *absolute value* of a number makes the negative sign "go away". Here are some examples:

$|-7| \rightarrow 7$
$|-4| \rightarrow 4$
$|5| \rightarrow 5$

Furthermore, we can solve equations with absolute values in them, such as:
$|x| = 5 \rightarrow x = 5, -5$
$|x+3| = 7 \rightarrow x = 4 \text{ or } -10$

Solve.
1) $|x - 4| = 9$
2) $|x + 8| = 12$
3) $|x - 3| + 2 = 6$

$|x-3| = 4$

Making Perfect Squares

The trinomial $x^2 - 10x + 25$ is called a *perfect square* because it happens to factor to a squared binomial: $(x-5)^2$.

Fill in the blank in order to create a perfect square trinomial.

Example: $x^2 + 8x +$ ___
Solution: 16 goes in the blank because $x^2+8x+16$ factors to $(x+4)^2$.

4) $x^2 + 6x +$ ___
5) $x^2 + 14x +$ ___
6) $x^2 - 12x +$ ___
7) $x^2 +$ ___ $+ 100$

Greek Geometric Puzzles

8) A rectangle has a length of 8 inches and a height equal to the length of the side of a square. Find the length of the side of the square such that the sum of the areas of the two figures is 65 square inches.

Homework

Complete the Square.
Fill in the blank in order to create a perfect square trinomial.

9) $x^2 + 4x +$ ___
10) $x^2 - 18x +$ ___
11) $x^2 - 2x +$ ___
12) $x^2 + 5x +$ ___
13) $x^2 +$ ___ $+ 36$

— The Quadratic Formula —

Solve.
14) $|3x + 5| = 2$
15) $|3x| + 5 = 2$
16) $|x + 5| = 1$
17) $|x - 3| = 7$
18) $|x| - 3 = 7$
19) $|½x - 5| + 3 = 11$

Word Problems.
20) The sum of two numbers is 32. The larger number is 12 more than twice the smaller number. Find the numbers.
21) The sum of the two numbers is 13, and the difference of their squares is 39. Find the numbers.

Problem Set #2

Group Work

1) $|x|$ is equivalent to which of the following:
   a) $5x \div 5$
   b) $\sqrt{x^2}$ ⟵ (circled)
   c) $8 + x - 8$
   d) All of the above
   e) None of the above

2) Given your answer to the previous problem (make sure that it is correct!), *fix the mistake* in solving this problem:
$$(x + 4)^2 = 49$$
$$\sqrt{(x+4)^2} = \sqrt{49}$$
$$x + 4 = 7$$
$$x = 7 - 4$$
$$x = 3$$

3) Solve by getting the squared term alone and then square rooting both sides. (Use the same method as shown above.)
   a) $(x - 3)^2 = 36$
   b) $x^2 - 4 = 5$

Greek Geometric Puzzles

4) A rectangle has a length of 10 inches and a height equal to the length of the side of a square. Find the side of the square such that the square has an area that is 24 square inches greater than the area of the rectangle.

5) A rectangle has a length of 10 inches and a height equal to the length of the side of a square. Find the side of the square such that the rectangle has an area that is 24 square inches greater than the square.

Homework

Solve by getting the squared term alone and then square rooting both sides.
6) $(x + 3)^2 = 16$
7) $(x - 7)^2 + 2 = 11$
8) $x^2 + 3 = 28$
9) $5x^2 - 3x + 25 = 3x(2x - 1)$

— The Quadratic Formula —
Problem Set #3

Group Work
Greek Geometric Puzzles

1) A rectangle has a length of 6 inches and a height equal to the length of the side of a square. Find the side of the square such that the square has an area that is 55 square inches greater than the rectangle.

2) A rectangle has a length of 8 inches and a height equal to the length of the side of a square. Find the side of the square such that the rectangle has an area that is 12 square inches greater than the square.

3) A rectangle has a length of 6 inches and a height equal to the side of a square. Find the side of the square such that the sum of the areas of the two figures is 20 square inches.

Solving a Quadratic Equation by Completing the Square
Study the following example and make sure that you understand it well.
Example: $x^2 + 10x - 24 = 0$
$x^2 + 10x = 24$
$x^2 + 10x + 25 = 24 + 25$
$(x + 5)^2 = 49$
$\sqrt{(x+5)^2} = \sqrt{49}$
$|x+5| = 7$
$x+5 = \pm 7$
$x = -5 \pm 7$
$x = -12, 2$

4) It would have been easier to solve the above example by factoring. Why do you think this new method is important?

Solve by Completing the Square (as shown above).

5) $x^2 + 8x + 12 = 0$

Homework
Solve by getting the squared term alone and then square rooting both sides, as was done on the previous set.

6) $(x - 2)^2 = 100$

7) $(x + 9)^2 = 1$

8) $(x + 5)^2 = 7$

9) $(x - 3)^2 = -4$

10) $(x + \frac{3}{8})^2 = \frac{9}{4}$

11) $(x - \frac{1}{3})^2 = \frac{5}{9}$

Solve by completing the square.

12) $x^2 - 6x + 5 = 0$

13) $x^2 + 4x - 21 = 0$

— The Quadratic Formula —

Problem Set #4

Homework
Solve by completing the square.

1) $x^2 + 8x - 20 = 0$
2) $x^2 - 4x = 32$
3) $x^2 + 6x = 3$
4) $x^2 + 3x - 28 = 0$
5) $x^2 - 7x + 11 = 0$

Solve for x in terms of the other variables or constants.

6) $3x - 2n = 7$
7) $5b - 2x = e$
8) $ax + b = c$
9) $bx - cd = b$

Word Problems

10) Seven times a smaller number is 4 more than twice a larger, and the sum of the two numbers is 16. What are the two numbers?

11) Jim's salary is $2/3$ of Alice's. Together they earn $600 per week. How much does each one make?

12) The difference of two numbers is 5. The sum of their squares is 233. Find the two numbers.

13) If the cost of a shirt is $4 more than the cost of a hat, and if 7 of those shirts cost the same as 9 of the hats, then what is the cost of the shirt?

14) Jane's quiz scores (out of 10 points and each worth 15%) were 6, 8, 9, and 4. Her final exam score (worth 40%) was an A (assume 95%). What is her final average?

15) Cathy is ten years older than twice Ben's age. In two years she will be three times his age. How old is Cathy now?

— The Quadratic Formula —

Problem Set #5
Classroom Discussion

Around 825AD, Mohammed ib'n Musa Al-Khwarizmi wrote *Hisab al-jabr wal-muqabala*. Most of the book focuses on arithmetic, measurement, business math and inheritance problems. But it is the first chapter, titled *On Calculating by Completion and Reduction*, that makes the book famous as the beginning of the formal study of algebra. We will now begin reading that first chapter.

On Calculating by Completion and Reduction

INTRODUCTION

When I considered what people generally want in calculating, I found that it is always a number. I also observed that every number is composed of units, and that any number may be divided into units.

Furthermore, I observed that the numbers which are required when calculating by completion and reduction are of three kinds, namely: roots, squares, and simple numbers[1]. Of these, a root is any quantity which is to be multiplied by a number greater than unity, or by a fraction less than unity. A square is that which results from the multiplication of the root by itself. A simple number [henceforth called only "number"] is any number which may be produced without any reference to a root or a square.

Of these three forms, then, two may be equal to each other, for example: squares equal to roots, squares equal to numbers, and roots equal to numbers[2].

Section I. CONCERNING SQUARES EQUAL TO ROOTS

The following is an example of squares equal to roots: "A square is equal to five roots". The root of the square then is five, and twenty-five forms its square, which is indeed equal to five times its root.

Another example: "One-third of a square equals four roots." Then the whole square is equal to 12 roots. So the square is 144, and its root is 12. Another such example: "Five squares equals ten roots." Therefore one square equals two roots. So the root of the square is two, and four represents the square.

In this manner, that which involves more than one square, or is less than one square, is reduced to one square. Likewise, the same is

---

[1] The term "roots" stands for multiples of the unknown, our x; the term "squares" stands for multiples of our x^2; "numbers" are constants.
[2] In our modern notation, this is $x^2 = bx$, $x^2 = c$, $x = c$.

done with the roots; that is to say, they are reduced in the same proportion as the squares.

Section II. CONCERNING SQUARES EQUAL TO NUMBERS

The following is an example of squares equal to numbers: "A square is equal to nine." Then nine is the square and three is the root. Another example: "Five squares equal 80." Therefore one square is equal to one-fifth of 80, which is 16. Or, to take another example: "Half of a square equals 18." Then the whole square equals 36, and its root is six.

Thus any multiple of a square can be reduced to one square. If there is only a fractional part of a square, you multiply it in order to create a whole square. Whatever you do, you must do the same with the number.

Section III. CONCERNING ROOTS EQUAL TO NUMBERS

The following is an example of roots equal to numbers: "A root is equal to three." Then the root is three and the square is nine. Another example: "Four roots equal 20." Therefore one root is five, and the square is 25. Still another example: "Half a root is equal to ten." Then the whole root is 20 and the square is 400.

[In addition to the three above cases] I have found that these same three elements can produce three compound cases, which are:

    Squares and roots equal to numbers,

    Squares and numbers equal to roots, and

    Roots and numbers equal to squares.

[These three cases are variations of a quadratic equation. Sections IV, V, and VI give methods for solving each of these cases. Only section IV, which deals with the first case, is covered in this workbook.]

Problem Set #6

Classroom Discussion

In section IV Al-Khwarizmi solves $x^2 + 10x = 39$.

Section IV. CONCERNING SQUARES AND ROOTS EQUAL TO NUMBERS

The following is an example of squares and roots equal to numbers: "A square and ten roots are equal to 39." The question here is: "What must the square be such that when it is combined with ten of its own roots, it will amount to a total of 39?" To solve this, you take half the number of roots, which in this case gives us five. Then you multiply this by itself to get 25, and then add that result to 39, which gives us 64. Now take the root of this, which is eight, and subtract from it half the number of roots, resulting in three. This is the root of the square which you sought; the square itself is then nine.

— The Quadratic Formula —

This method is the same when you are given a number of squares. You simply reduce them to a single square, and in the same proportion you reduce the roots and simple numbers that are connected with them.

For example: "Two squares and ten roots equal 48." The question therefore is: "What must the amount of the two squares be such that when they are summed up and then combined with ten times their root, the result will be a total of 48?" First of all it is necessary that the two squares be reduced to one. So we take half of everything mentioned in the statement. It is the same now as if the original question had been: "A square and five roots equal 24", which means: "What must the amount of a square be such that when it is combined with five times its root, the result will be a total of 24?" To solve this, we halve the number of roots, which gives us 2½, and multiply that by itself, giving 6¼. To this we add 24, which yields a sum of 30¼, and then take the root of this, which is 5½. Subtracting half the number of roots, which is 2½, from this makes a remainder of three. This is the root of the square, and the square itself is nine.

$x^2 + bx = c$

Group Work
Al-Khwarizmi's Formula
1) Essentially, Al-Khwarizmi has just given a formula for solving quadratic equations. Consider the number of roots (i.e. the coefficient of the x term) to be b, and the constant to be c. Now reread Section IV in order to derive a formula for x given in terms of b and c.

Homework
Solve by completing the square.
2) $x^2 + 12x - 28 = 0$
3) $x^2 - 2x - 7 = 0$
4) $x^2 - 2x + 7 = 0$
5) $x^2 + 5x - 50 = 0$
6) $x^2 - 9x + 5 = 0$
7) $2x^2 + 11x + 12 = 0$

Solve for x in terms of the other variables or constants.
8) $ax - b^2 = d$
9) $a(x + b) = c$

Problem Set #7

Classroom Discussion In section V Al-Khwarizmi gives a solution for solving *Squares and Numbers Equal to Roots* ($10x = x^2 + 21$). Section VI gives a solution for solving *Roots and Numbers Equal to Squares* ($3x + 4 = x^2$). We are skipping over these two sections. We will now pick up with the last paragraph of section VI, which reads as follows:

I have now explained the six types of equations, which I first mentioned at the beginning of this book. I have taught how the first

The Quadratic Formula

three must be solved; with these, it was not required that the roots be halved. And I showed how, with the other three, halving the roots is necessary. I now think it is necessary to explain the reason for halving.

Section VII. A DEMONSTRATION OF THE CASE
"A Square and Ten Roots Equal 39."

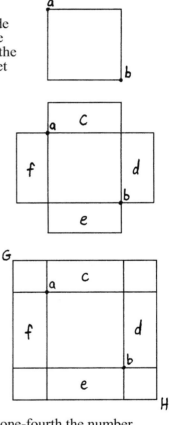

First, we construct a square ab of unknown sides. This square represents the square which, together with its root, you wish to find. Any side of this square represents one of the roots that we wish to know. We will now take one-fourth of the number of roots, namely one-fourth of ten, to get 2½. Combining this with the side of the square gives us four new rectangles (c, d, e, f), which we will place onto the sides of the square [as shown in the middle drawing].

We now have a new, larger square except that small square pieces are missing from its four corners. These four corners each have an area of 2½ times 2½. When we add these four corners to our figure [as shown in the lower drawing], we have increased the area by four times the square of 2½, which is 25.

From the original statement we know that the square ab combined with the four rectangles, which together represent ten roots, must be equal to a total of 39. To this we add 25 (the area of the four small corners) to get a total of 64, which is the area of the great square GH. One side of this great square must then be eight. If we subtract twice a fourth of ten, which is five, from this eight then we get three – the root of the square which we sought.

It must be observed that here we have taken one-fourth the number of roots, multiplied that result by itself, and then multiplied that by four, which is the equivalent of taking half the number of roots and then multiplying that by itself [which is what was done in section IV].

— The Quadratic Formula —

Group Work

1) Express the statement
   A square and 8 roots equal 65
   a) As a modern algebra equation.
   b) As a Greek geometric puzzle.

2) Solve problem #8 from Set #1 using...
   a) Al-Khwarizmi's geometrical method (as in Section VII).
   b) Al-Khwarizmi's formula (as in Section IV).
   c) Completing the square.

Problem Set #8

Note: From this point forward, all answers involving irrational numbers should be given both as an exact (perhaps irrational) number, and as a decimal approximation.

Group Work

1) Solve problem #3 from Set #3 using...
   a) Al-Khwarizmi's geometrical method.
   b) Al-Khwarizmi's formula.
   c) The method of completing the square.

2) Consider problem #2 from Set #3. Solve it using the easiest of the above three methods.

3) *Challenge!* Consider problem #1 from Set #3.
   a) How would al-Khwarizmi have stated this problem?
   b) Show how he might have solved it geometrically.

   c) Give a formula that he might have given for solving this problem.

Solve by completing the square.

4) $4x^2 - 21x + 5 = 0$
5) $x^2 + bx + c = 0$ (Your answer to this should be the same as from a problem on an earlier set. Which one is it?)

Homework

Solve by completing the square.

6) $x^2 - 4x - 60 = 0$
7) $x^2 + 3x - 5 = 0$
8) $x^2 + 3x + 5 = 0$
9) $5x^2 + 13x - 6 = 0$
10) $3x^2 + 13x + 5 = 0$

Problem Set #9

Group Work

Solve by Completing the Square.

1) $3x^2 + 11x + 5 = 0$
   (Leave your answer in square root form.)

2) $ax^2 + bx + c = 0$
   (Your answer is the Quadratic Formula!)

— The Quadratic Formula —

Homework
Solve by completing the square.
3) $x^2 + x - 5 = 0$
4) $6x^2 - 19x + 10 = 0$
5) $3x^2 + 4x + 5 = 0$
6) $3x^2 + 4x - 5 = 0$

Word Problems
7) The length of a rectangle is 3m more than the width. What are the dimensions if the perimeter is 15m?

8) Find the width of a rectangle if twice the width is six feet more than the length, and the area is 80 ft^2.

9) A rectangle has a length of 18 inches and a height equal to the length of the side of a square. Find the side of the square such that the rectangle has an area that is 80 square inches greater than the square.

Problem Set #10

Group Work
Solve the equation using each of three methods:
  a) Factoring.
  b) Completing the Square.
  c) The Quadratic Formula.
1) $x^2 + 9x + 20 = 0$
2) $6x^2 + 7x - 10 = 0$

Homework
Solve by using each of the three methods:
3) $x^2 - 6x - 16 = 0$
Solve by quadratic formula:
4) $3x^2 - 8x + 4 = 0$

Solve by the easiest method:
5) $x^2 + 9x + 14 = 0$
6) $x^2 + 5x - 11 = 0$
7) $3x^2 + 10x + 8 = 0$
8) $5x^2 + 7x - 10 = 0$
9) $x^2 + 6x = 3$

Word Problems
10) The length of a rectangle is 6 inches less than four times the width. If the perimeter is 23 inches, then what are the dimensions?

Problem Set #11

Homework
1) Give the Quadratic Formula.
2) The Quadratic Formula is the solution to what equation?
3) Give the proof of the quadratic formula.

Solve the equation using each of the three methods (as stated on the previous set):
4) $x^2 - 7x + 12 = 0$

The Quadratic Formula

Solve by using the easiest method:
5) $x^2 + 8x + 5 = 0$
6) $x^2 + 2x - 35 = 0$
7) $x^2 + 2x + 3 = 0$
8) $4x^2 + x - 3 = 0$
9) $3x^2 + 5x + 2 = 0$
10) $7x^2 + 8x - 3 = 0$

Word Problems
The following three problems are the same as the first three problems in set#3, except that the numbers have been changed.

11) A rectangle has a length of 10 inches and a height equal to the length of the side of a square. Find the side of the square such that the square has an area that is 56 square inches greater than the rectangle.

12) A rectangle has a length of 8 inches and a height equal to the length of the side of a square. Find the side of the square such that the rectangle has an area that is 15 square inches greater than the square.

13) A rectangle has a length of 7 inches and a height equal to the length of the side of a square. Find the side of the square such that the sum of the areas of the two figures is 50 square inches.

14) What are the dimensions of a rectangular garden that has a perimeter of 66 ft and an area of 216 ft²?

15) Wendy has nickels, dimes and quarters, 18 coins in all, worth a total of two dollars. How many of each coin are there if there are twice as many dimes as nickels?

Problem Set #12

Homework
1) Give the Quadratic Formula.
2) The Quadratic Formula is the solution to what equation?

Solve.
3) $x^2 + 6x + 2 = 0$
4) $3x^2 + 5x - 3 = 0$
5) $2x^2 + 3x - 5 = 0$
6) $2x^2 + 3x = 3(x+7)$
7) $\frac{4x}{x-2} = \frac{x-5}{x-3}$
8) $\frac{2}{x-1} = \frac{3}{x-3} + \frac{2}{x-4}$
9) $(x+4)(x-5) = 2x^2 - 4x - 48$
10) $(2x+3)(2x-3) = -x - 6$
11) $(x-3)^2 = 3x^2 + 4x + 12$

Find the common solution.
12) $x + 4y = 3$
    $y^2 - 2x = 10$

13) What can be said about the relationship of a, b, c in the case that a quadratic equation has no solution?

The Quadratic Formula

Word Problems

14) A rectangle has a length of 10 inches and a height equal to the length of the side of a square. Find the side of the square such that the sum of the areas of the two figures is 20 square inches.

15) The sum of two numbers is 15. The square of one of them plus twice the other is 54. What are the two numbers?

16) A rectangular garden plot runs along the side of a building and is surrounded on the other three sides by a fence. Find the dimensions of the plot if the total length of the fence is 17m and the area of the plot is 35m^2.

17) A 10m tall wooden pole snaps just below its middle. Remaining connected, the top portion falls over, and touches the ground 3m from the base. How far from the base did the break occur?

Problem Set #13

Homework
Solve.

1) $x^2 = x + 1$
2) $(x + 5)^2 = 3x^2 - 23$
3) $(x + 6)(x - 4) = x^2$
4) $(x + 6)(x - 4) = 2x^2$
5) $10x^2 + 40x + 20 = 0$
6) $5 - 2(3x - 4) = 3x^2 - 6x$
7) $\frac{3}{x-3} - \frac{3x-1}{x} = \frac{9}{x^2-3x}$
8) $2x^2(x-5) + 16 = (3x-4)^2$

Find the common solution.

9) $2x + y = 7$
   $x^2 + 3y = 13$

Word Problems

10) One number is three less than twice another number. The sum of their squares is 94¼. Find the two numbers.

11) Karen is ⅔ as old as Bill. Three years ago, the product of their ages was 273. How old is Karen now?

12) A rectangular piece of cardboard is twice as long as it is wide. If squares measuring 3 inches on a side are cut off each of the corners of the cardboard, then the four sides can be folded up to form a open-topped box that has a volume of 300 cubic inches. How long is the longest edge of the box?

13) (From Euclid's *The Elements,* Book II, Th. 11)

a) Where can you cut a 10cm-long straight line such that the rectangle formed by the whole line and one of the segments is equal to the square on the remaining segment?

b) What is the ratio (in decimal form) of the lengths of the two segments found above?

Logarithms – Part I

This unit is a brief introduction to an important topic. In the next two workbooks, logarithms are revisited (and covered in more depth) in *Logarithms, Part II* and *Logarithms, Part III*.

Problem Set #1

Simplify.
1) $(x^3)^2$
2) $x^3 \cdot x^2$
3) $(x^6)^4$
4) $x^6 \cdot x^4$
5) $x^3 + x^2$
6) $x^4 + x^4$
7) $(4x^3)^5$
8) $3x^4 \cdot 7x^5$
9) $(3x^5)^4$
10) $2x^3 \cdot 3x^9$
11) $4x^3 + 2x^5$
12) $2x^3 + 5x^3$
13) $3x^3 - x^3$

Give a simplified answer that has no negative exponents.
14) $x^{-4}x^7$
15) $\dfrac{x^{-4}}{x^3}$
16) $5x^{-4}$
17) $3x^2 y^{-5}$
18) $3y^{-5}x^2$
19) $3y^{-5} + 4x^2$
20) $(\tfrac{2}{3})^4$
21) $(\tfrac{2}{3})^{-4}$
22) $(\tfrac{2}{3})^{-1}$
23) $(\tfrac{2}{3})^0$
24) $(x^{-4})^5$

25) $(3x^{-3})^2$
26) $(9x^{-4}y^3)^4$
27) $(6x^{-3}y^7)^3$
28) $(6x^{-6}y^2)^3$
29) $(4x^{-6}y^2)^{-4}$
30) $(8x^6 y^{-5})^{-3}$
31) $\left(\dfrac{x^{-3}}{y^2}\right)^3$
32) $\dfrac{15x^{-4}y^{-3}}{6x^{-7}y^0}$
33) $\dfrac{8x^{-4}y^7}{6x^3 y^3}$
34) $\left(\dfrac{8x^{-4}y^7}{6x^3 y^3}\right)^{-1}$
35) $\left(\dfrac{8x^{-4}y^7}{6x^3 y^3}\right)^2$
36) $\left(\dfrac{8x^{-4}y^7}{6x^3 y^3}\right)^0$

Calculate each. Use the *Power and Base Tables* (on the next page), if needed. Leave square roots in simplified form.
37) 8^2
38) 8^{-2}
39) 6^0
40) 6^{-1}

41) 6475^{-1}
42) 7384^0
43) 7^{-3}
44) $\log_3 9$
45) $\log_2 16$
46) $\log_5 25$
47) $\log_5 625$
48) $\log_4 64$
49) $\log_2 64$
50) $\log_8 64$
51) $\log_2 1024$
52) $\log_9 531441$
53) $\log_7 16807$
54) $\log_2 2$
55) $\log_2 1$
56) $\log_{57} 57$
57) $\log_{29} 1$
58) Explain why each one is true:
  a) $2^{10} = 4^5$
  b) $3^{10} = 9^5$
  c) $2^9 = 8^3$
  d) $4^6 = 8^4$
59) A rectangle's length is 4 more than three times its width. Find its length if the perimeter is 52.

— Logarithms —

Power and Base Tables

2nd Power

N	N^2
1	1
2	4
3	9
4	16
5	25
6	36
7	49
8	64
9	81
10	100

3rd Power

N	N^3
1	1
2	8
3	27
4	64
5	125
6	216
7	343
8	512
9	729
10	1000

4th Power

N	N^4
1	1
2	16
3	81
4	256
5	625
6	1296
7	2401
8	4096
9	6561
10	10000

5th Power

N	N^5
1	1
2	32
3	243
4	1024
5	3125
6	7776
7	16807
8	32768
9	59049
10	100000

Base 2

N	2^N
1	2
2	4
3	8
4	16
5	32
6	64
7	128
8	256
9	512
10	1024

Base 3

N	3^N
1	3
2	9
3	27
4	81
5	243
6	729
7	2187
8	6561
9	19683
10	59049

Base 4

N	4^N
1	4
2	16
3	64
4	256
5	1024
6	4096
7	16384
8	65536

Base 5

N	5^N
1	5
2	25
3	125
4	625
5	3125
6	15625
7	78125

Base 6

N	6^N
1	6
2	36
3	216
4	1296
5	7776
6	46656

Base 7

N	7^N
1	7
2	49
3	343
4	2401
5	16807
6	117649

Base 8

N	8^N
1	8
2	64
3	512
4	4096
5	32768
6	262144

Base 9

N	9^N
1	9
2	81
3	729
4	6561
5	59049
6	531441

— Logarithms —

Problem Set #2

Calculate each. Use the *Power and Base Tables*, if needed. Leave square roots in simplified form.

1) 5^3
2) 5^{-3}
3) 5^0
4) 5^1
5) 5^{-1}
6) 1875^0
7) 1875^1
8) 1875^{-1}
9) 3^6
10) $(-3)^6$
11) 3^7
12) $(-3)^7$
13) $49^{1/2}$
14) $250,000^{1/2}$
15) $3600^{1/2}$
16) $90,000,000,000^{1/2}$
17) $8^{1/3}$
18) $8000^{1/3}$
19) $27^{1/3}$
20) $27,000,000^{1/3}$
21) $64^{1/2}$
22) $64^{1/3}$
23) $64,000,000^{1/2}$
24) $64,000,000^{1/3}$
25) $256^{1/2}$
26) $256^{1/4}$
27) $256^{1/8}$
28) $25,600,000,000^{1/2}$
29) $25,600,000,000^{1/4}$
30) $25,600,000,000^{1/8}$
31) 4^2
32) 4^{-2}
33) $4^{1/2}$
34) $4^{-1/2}$
35) 7^2
36) 7^{-2}
37) $7^{1/2}$
38) $7^{-1/2}$
39) $90,000^2$
40) $90,000^{-2}$
41) $90,000^{1/2}$
42) $90,000^{-1/2}$
43) 8^3
44) 8^{-3}
45) $8^{1/3}$
46) $8^{-1/3}$
47) 8000^3
48) 8000^{-3}
49) $8000^{1/3}$
50) $8000^{-1/3}$
51) 25^2
52) 25^{-2}
53) $25^{1/2}$
54) $25^{-1/2}$
55) $6561^{1/4}$
56) $6561^{-1/4}$
57) $6561^{1/8}$
58) $6561^{-1/8}$
59) $\log_2 8$
60) $\log_2 16$
61) $\log_3 9$
62) $\log_7 49$
63) $\log_2 1024$
64) $\log_3 81$
65) $\log_4 16$
66) $\log_4 64$
67) $\log_8 64$
68) $\log_2 64$
69) $\log_{64} 64$
70) $\log_8 (1/64)$
71) $\log_5 25$
72) $\log_5 (1/25)$
73) $\log_{70} 4900$
74) $\log_7 (1/49)$
75) $\log_2 8$
76) $\log_2 (1/8)$
77) $\log_{25} 5$
78) The length of a rectangle is four times its width. Find the length if the area is 49m².

— Logarithms —

Problem Set #3

Give a simplified answer that has no negative exponents.

1) $(3/4)^{-1}$
2) $(3/4)^0$
3) $(3/4)^3$
4) $(3/4)^{-3}$
5) $(3x^{-3}y^4)^{-3}$
6) $\left(\dfrac{x^3}{y^2}\right)^{-1}$
7) $\left(\dfrac{x^3}{y^2}\right)^{-3}$
8) $\dfrac{18x^4y^{-3}}{24x^{-6}y^{-5}}$
9) $\left(\dfrac{10x^{-2}y^{-5}}{6x^{-6}y^3}\right)^{-3}$

Calculate each. You may use the *Power and Base Tables*.

10) $16^{1/2}$
11) $16^{1/4}$
12) $1{,}600{,}000{,}000^{1/2}$
13) $1{,}600{,}000{,}000^{1/4}$
14) $1{,}600{,}000{,}000^{-1/2}$
15) $1{,}600{,}000{,}000^{-1/4}$
16) $1{,}600{,}000{,}000^{3/4}$
17) $1{,}600{,}000{,}000^{3/2}$
18) $1{,}600{,}000{,}000^{3/2}$
19) $125^{2/3}$
20) $125^{1/3}$
21) $64^{-4/3}$
22) $64^{-1/2}$
23) $64^{5/2}$
24) $64^{-5/6}$

25) $64^{-2/3}$
26) 900^2
27) 900^{-2}
28) $900^{1/2}$
29) $900^{-1/2}$
30) $\log_3 27$
31) $\log_{300} 27{,}000{,}000$
32) $\log_8 512$
33) $\log_8 \left(\dfrac{1}{512}\right)$
34) $\log_8 2$
35) $\log_8 (1/2)$
36) $\log_9 81$
37) $\log_9 \left(\dfrac{1}{81}\right)$
38) $\log_9 3$
39) $\log_9 (1/3)$
40) $\log_9 27$

41) Write in scientific notation:
   a) 673,000,000
   b) 70,000,000,000
   c) 0.00253

42) Write in standard form:
   a) $7.5 \cdot 10^7$
   b) $8.04 \cdot 10^{-5}$

43) Change to exponent form:
   Example: $\log_2 8 = 3$
   Solution: $2^3 = 8$
   a) $\log_5 625 = 4$
   b) $\log_{10} 0.001 = -3$
   c) $\log_8 (1/4) = -2/3$

44) Change to log form:
   a) $6^3 = 216$
   b) $5^{-2} = 1/25$
   c) $16^{3/4} = 8$

— Logarithms —

Problem Set #4

Calculate each. You may use the *Power and Base Tables*.
** Indicates answers should be given in scientific notation.

1) 400^2
2) 400^{-2}
3) $400^{1/2}$
4) $400^{-1/2}$
5) **$400^{5/2}$
6) $400^{-3/2}$
7) **$8{,}000{,}000^3$
8) $8{,}000{,}000^{-3}$
9) **$8{,}000{,}000^{1/3}$
10) $8{,}000{,}000^{-1/3}$
11) **$8{,}000{,}000^{2/3}$
12) **$8{,}000{,}000^{-2/3}$
13) $1{,}000{,}000{,}000{,}000^{1/4}$
14) $729^{1/6}$
15) $729^{5/6}$
16) $729^{-5/6}$
17) $729^{1/3}$
18) $729^{-1/3}$
19) $729^{2/3}$
20) $729^{-2/3}$
21) $\log_4 16$
22) $\log_{40} 1600$
23) $\log_{40} 64000$
24) $\log_4 (1/4)$
25) $\log_4 1$
26) $\log_4 2$
27) $\log_4 (\frac{1}{16})$
28) $\log_4 (-1/2)$
29) $\log_{25} 625$
30) $\log_{25} (\frac{1}{625})$
31) $\log_{25} (\frac{1}{5})$
32) $\log_{25} 125$
33) $\log_{25} (\frac{1}{125})$
34) $\log_{100} 1000000$
35) $\log_{100} 10$
36) $\log_{100} 1000$
37) $\log_{100} 0.1$
38) $\log_{100} 0.01$
39) $\log_{100} 0.001$
40) $\log_{27} 81$
41) $\log_6 (-36)$
42) $\log_{81} 3$
43) $\log_8 (\frac{1}{256})$
44) Write in scientific notation:
  a) 90,700,000
  b) 730,000,000,000
  c) 0.3
45) Write in standard form:
  a) $7.08 \cdot 10^4$
  b) $8 \cdot 10^{-8}$
46) Change to exponent form:
  Example: $\log_2 8 = 3$
  Solution: $2^3 = 8$
  a) $\log_{10} 100000 = 5$
  b) $\log_4 (\frac{1}{64}) = -3$
  c) $\log_3 4x = 5$
47) Change to log form:
  a) $7^3 = 343$
  b) $8^{-3} = \frac{1}{512}$
  c) $9^{4x+7} = 285$
48) A square sheet of paper has 2 cm cut off its side and 3 cm cut off its top, thereby losing 94 cm² of area. Find the length of the side of the original square.

— Logarithms —

Problem Set #5

Calculate each. Use the *Power and Base Tables*, if needed. ** Indicates that answers should be given in scientific notation.

1) 36^2
2) $36^{1/2}$
3) 36^{-2}
4) $36^{-1/2}$
5) $1024^{1/5}$
6) $1024^{-1/5}$
7) **$16{,}000{,}000{,}000{,}000^2$
8) **$16{,}000{,}000{,}000{,}000^{1/4}$
9) **$16{,}000{,}000{,}000{,}000^{5/2}$
10) **$16{,}000{,}000{,}000{,}000^{-3/4}$
11) $\log_9 729$
12) $\log_9 \left(\frac{1}{729}\right)$
13) $\log_6 1296$
14) $\log_3 \left(\frac{1}{729}\right)$
15) $\log_9 \left(\frac{1}{3}\right)$
16) $\log_3 \left(\frac{1}{9}\right)$
17) $\log_8 16$
18) $\log_{16} 8$
19) $\log_5 (-25)$
20) $\log_{37} \left(\frac{1}{37}\right)$
21) $\log_{25} \left(\frac{1}{125}\right)$
22) $\log_{81} \left(\frac{1}{27}\right)$

Solve. It may help to rewrite the equation in a different form (e.g. exponential or log form). Use a calculator only if necessary.

23) $5^4 = x$
24) $5^X = \frac{1}{125}$
25) $x^4 = 625$
26) $x^3 = 30$
27) $5 \cdot 4^{3x} = 5120$
28) $\log_x 64 = 6$
29) $\log_x 6 = 2$
30) $\log_4 x = 3$
31) $\log_2 32 = x$
32) $\log_5 5x = 3$
33) $4 + 5 \log_3(2x+7) = 24$
34) $\frac{1}{9} 6^{3x-5} - 6 = 18$
35) $8 + 3 \log_5 (2x-7) = 17$

36) A rectangular garden is twice as long as it is wide. By increasing the length and width of the garden by 2 feet each, its area is increased by 40 ft². Find the dimensions of the original garden.

Possibility & Probability – Part I

This unit is intended to review and deepen material that has been introduced in main lesson. This topic is covered one more time in eleventh grade with *Possibility & Probability, Part II*.

Problem Set #1

Section A

1) Bob's Bikes makes bikes with 2 types of frames, 3 handle bar styles, and in colors red, yellow, green, black, or white. How many different bikes can they make?

2) Paul's Pizza offers 3 choices of salad, 20 kinds of pizza, and 4 different desserts. How many different 3-course meals can be ordered?

3) How many 7-digit phone numbers are possible? (The first digit cannot be 0 or 1.)

4) In how many ways can 8 people be lined up in a row?

5) A license plate consists of 3 letters followed by 3 digits (e.g., XBB022). How many different plates could be issued?

6) How many ways can four different roles in a play be assigned from a group of 14 actors?

7) In how many ways can a president and a secretary be chosen from a group of 6 people?

8) In a 6-horse race…
   a) how many different orders of finishing are there?
   b) how many possibilities are there for the first 3 places?

9) A character can be either a letter or a digit. (Thus, there are 36 different characters.)
   a) How many possible three-character codes are there?
   b) How many possible three-character codes have different characters and a digit as the first character?

Section B

10) How many license plates are possible that have 2 digits and 2 letters (in any order)?

11) Using the letters of the word EQUATION, how many 4-letter words (which don't have to spell anything) can be formed (without repetition)…
    a) that start with T?
    b) that start and end with a consonant?
    c) that have only one vowel?
    d) with all the vowels positioned furthest to the right?

— Possibility & Probability —

Problem Set #2

Section A

1) Calculate
   (a) 6! (b) $_5P_5$ (c) $_5P_3$ (d) $_5P_1$
   (e) $_5C_5$ (f) $_5C_3$ (g) $_5C_1$ (h) $_5C_0$

2) John has 4 ties, 6 shirts, and 3 pairs of pants. How many different outfits can he wear? Assume that he wears one of each kind of article.

3) How many 7-digit telephone numbers can be created if the first digit must be 4, the second must be 7, and the third must be 5 or 6?

4) In how many ways…
   a) can the letters of the word "FRIDAY" be arranged?
   b) can the letters of the word "LESSON" be arranged?

5) How many three-digit numbers are there that use only the digits 0,1,2,3,4? (A number can't start with 0.)

6) In how many ways can first, second, and third prizes be awarded in a competition in which there are 14 entries?

7) From a group of 14 people, how many ways can a committee of 3 be chosen?

8) There are 13 different colored crayons. How many different ways can you choose four of them?

9) You have a pack of 13 different colored pencils. You wish to color these 4 squares in 4 different colors. In how many different ways can you do it?

10) In how many ways…
    a) can a committee of 3 be selected from a group of 8?
    b) can a committee of 5 be selected from a group of 8?

11) One marble is drawn at random from a bag containing 4 white, 5 red, and 6 green marbles. Find the probability that...
    a) it is white.
    b) it is red or green.
    c) it is not white.

12) Three coins are tossed. What is the probability of getting…
    a) all heads?
    b) exactly 2 heads?

Section B

13) In a group of 10 people, each person shakes hands with everyone else once. How many handshakes are there?

14) There are 36 numbers on a combination lock. To open the lock you must know the three correct numbers (which are all different) in the correct order. How long would it take a thief to try all possible combinations if it takes 10 seconds to try each possibility?

— Possibility & Probability —

Problem Set #3

Section A

1) Gail is buying a certain model bike. She has a choice of four different colors, two kinds of handlebars, two kinds of tires, and three different pedals. How many different kinds of bikes are there?

2) How many 5-digit passwords can be created if the first digit must be a 3, the last digit must be odd, and you can't repeat a digit?

3) How many ways can you choose 29 things out of 30 without regard to order?

4) Suppose that a club consists of 8 women and 6 men. In how many ways can a president and a secretary be chosen if...
   a) there are no restrictions?
   b) the president is to be female and the secretary male?
   c) the president is to be male and the secretary female?
   d) both are to be female?
   e) the president and secretary are to be of opposite sex?

5) A class has 7 boys and 10 girls as members. How many different 6-person committees can be selected...
   a) from all the members?
   b) if there must be an equal numbers of boys and girls?

6) There are 6 students in a class. What is the probability that they will arrive to class on a given day in alphabetical order?

7) A box contains 13 cards numbered 1 through 13. Suppose one card is drawn from the box. Find the probability that...
   a) The number drawn is even.
   b) The number is greater than 9 or less than 3.

8) Use Pascal's triangle to expand...
   a) $(x+y)^4$
   b) $(x+2)^4$

9) Four coins are tossed. What is the probability of getting...
   a) all tails?
   b) exactly two heads?
   c) at least two heads?

10) How many ways are there to arrange the letters SSSEEEEPP?

Section B

11) How many different ways are there to arrange 8 identical blue chairs and 6 identical red chairs in a row?

12) Two dice are rolled. Find the probability that...
   a) The sum of the numbers showing on the dice is 10.
   b) The sum of the numbers showing on the dice is 8.
   c) Exactly one die shows a 4.

— Possibility & Probability —

Problem Set #4

Section A

1) In how many different ways can a 10-question multiple-choice test be answered if every question has A, B, C, or D as its options?

2) A committee of 4 is to be selected from a group of 3 seniors, 4 juniors, and 5 sophomores. In how many ways can it be done if
   a) there are no restrictions on the selection?
   b) the committee must have 2 sophomores, 1 junior, and 1 senior?
   c) the committee must have at least 3 sophomores?
   d) the committee must have at least 1 senior?

3) A single marble is drawn from a bag containing 3 red, 5 white, and 4 blue marbles. Find the probability that...
   a) A red marble is drawn.
   b) A red or blue marble is drawn.
   c) A blue or white marble is drawn.
   d) A red, white, or blue marble is drawn.

4) There are 5 multiple-choice questions on an exam, each with 4 possible answers. What is the probability of getting all 5 answers correct, if you guess randomly?

5) What is the probability of randomly, but correctly, guessing the top three finishers in an 8-horse race?

6) Two dice are rolled. What is the probability of getting…
   a) a 7?
   b) an 11?
   c) a 7 or an 11?

7) How many different…
   a) poker hands are possible? (Poker hands consist of 5 cards.)
   b) bridge hands are possible? (Bridge hands consist of 13 cards.)

8) One card is drawn at random from a 52-card deck. Find the probability that...
   a) It is an ace.
   b) It is a diamond.
   c) It is black.

Section B

9) Using the letters of the word "TENNESSEE"…
   a) How many different ways can the letters be arranged?
   b) How many different ways can the letters be arranged so that the 4 E's are in consecutive positions?

10) You and your friend are both in a group of 20 people, and 5 people are to be randomly selected to be on a committee. What is the probability that both you and your friend will be on the committee?

— Possibility & Probability —

Problem Set #5

Section A

1) How many 5-letter words (which don't have to spell anything) can be made using A, B, C, D, and E…
   a) if each letter may only be used once?
   b) if letters may repeat?

2) There are 10 applicants for three different job positions at a department store. How many ways are there to fill the three positions?

3) In how many ways can a student select 4 college courses from a set of 9 courses (that meet at different times)?

4) How many different arrangements of the word "MISSISSIPPI" are there?

5) In a class of 12 students, the teacher must choose a different student for each day of the coming school week to give a presentation. How many possible line-ups are there?

6) In a class of 12 students, the teacher must choose five students to go to a math party. How many possible choices are there for this group?

7) On a circle lie 10 points. How many chords (connecting lines) can be drawn between these points?

8) In how many ways can 3 red, 4 blue, and 2 green pens be distributed to 9 students seated in a row if each student receives one pen?

9) One card is drawn at random from a 52-card deck. Find the probability that…
   a) it is the king of clubs.
   b) it is a red queen.
   c) it is a 7 or 8.

10) A coin is tossed 5 times. Find the probability that…
    a) they will all be heads.
    b) exactly two will be tails.

11) Use Pascal's triangle to expand…
    a) $(x+y)^5$
    b) $(x+10)^5$

Section B

12) How many ways can all 16 white chess pieces be arranged in a row? (Each color has 8 pawns, 2 rooks, 2 bishops, 2 knights, 1 king, and 1 queen.)

13) How many ways…
    a) can we break 21 students into 7 groups of three?
    b) can they be broken into 3 groups of seven?

14) A coin is tossed 7 times. Find the probability that…
    a) At least 5 tosses come up tails.
    b) At least one toss comes up heads.

15) With a five-card poker hand, what is the probability of getting…
    a) a pair of aces and a pair of kings?
    b) two pairs of different kinds (e.g., two aces and two 5's)?

— Possibility & Probability —

Problem Set #6

Section A

1) How many positive odd integers less than 10,000 can be written using only the digits 3,4,7,8, and 0 (and allowing for repeat digits)?

2) How many different ways are there to rearrange the letters of "STATISTICS"?

3) On a restaurant's menu there are 8 main courses, and 5 desserts. How many ways are there...
   a) to order a main course and a dessert?
   b) to order three different main courses to be shared between friends?

4) A baseball team has 15 players, four of whom pitch. How many ways can the awards best pitcher, most valuable player, and most improved player be given if...
   a) any player can receive more than one award?
   b) any player can only receive at most one award?

5) Six numbers are drawn from a hat. What is the probability that the numbers will be drawn in either ascending or descending order?

6) Two cards are drawn from a 52-card deck. Find the probability that...
   a) both are hearts.
   b) neither is red.

Section B

7) A pizza parlor has 12 different possible toppings that could be put on your pizza. How many possible ways could you choose 3 toppings or fewer?

8) In how many ways can 9 (different) presents be distributed to 3 children if each is to receive 3 presents?

9) Consider the map shown here:

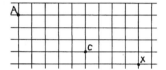

   a) How many different shortest routes (no backtracking, no cutting across blocks) are there from A to C?
   b) How many different shortest routes are there from A to X?

10) Four cards are drawn from a 52-card deck. Find the probability that you will pick a 6, 7, 8, and 9?

11) With a five-card poker hand, what is the probability of getting...
    a) no clubs.
    b) all cards from the same suit.
    c) only 5's and 6's.

Cartesian Geometry – Part I

Problem Set #1

In 1637, René Descartes published a book with the impressive title *Discourse on the Method of Rightly Conducting One's Reason and Searching for the Truth in the Sciences*, which today is seen as a work of major importance in the fields of philosophy and in general science. The book also included an appendix on geometry where he showed how his new method of conducting science could be used to develop a new way of solving geometric problems. Before Descartes, geometry and algebra were separate subjects.

What Descartes actually did was to take a geometry problem (the Pappus problem) and expressed it as an equation. Descartes' seed idea (and Fermat came up with similar ideas at about the same time) was then further developed over a period of time, and has been tremendously influential in the world of science and mathematics. Modern Cartesian geometry (which is also referred to as coordinate geometry or analytical geometry) allows us to take an equation and express it as geometry; *it allows us to visualize algebra.*

To graph an equation, we simply follow Descartes' words: "We may give any value we please to either x or y, and find the value of the other from the equation", in order to find several solutions. Then you plot the solutions to each equation on a Cartesian graph. For many equations, you have to carefully choose the values for x or y so that the points you plot fit reasonably on the graph.

Example: $y = \frac{3}{4}x - 2$
Solution: It is easiest, perhaps, to plug in multiples of 4 into x. (Although we could have instead chosen fairly random values for x.) This leads to the following table:

x	–4	0	4	8	12
y	–5	–2	1	4	7

Note that for every step of the four that the x values take, the y values take a step of three. (We will see later why this is important.) Lastly, we simply need to graph the solutions that we have found, and we see that the result is the straight line shown here:

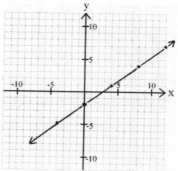

Cartesian Geometry – Part I

Graph each equation.
1) $x^2 + y^2 = 25$
2) $y = \frac{1}{2}x + 3$
3) $y = x^2 + 6x + 5$
4) $y = 2x^3 - 3x + 1$
5) $y = x^4 - 5x^2 + 4$

Problem Set #2

On the previous problem set we were able to graph equations making a table and then plotting points. While the method of making a table and plotting points can be reliable, it is time consuming and tedious.

You may have noticed on the last problem set that the equations without any exponents ended up having graphs that were straight lines. Such equations are called *linear equations*. Mathematicians are always searching for more efficient ways to do things; this problem set is focused on finding quicker ways to graph linear equations.

1) Graph each of the following on the same graph by making a table and then plotting points.

   a) $y = 2x + 1$
   b) $y = 2x - 3$
   c) $y = 2x + 4$
   d) $y = 2x - 6$

2) With the above equations, what does the number at the end of the equation tell you?

3) Graph each of the following on the same graph by making a table and then plotting points.

   a) $y = 2x + 1$
   b) $y = \frac{2}{5}x + 1$
   c) $y = \frac{5}{2}x + 1$
   d) $y = -\frac{5}{2}x + 1$
   e) $y = -\frac{3}{5}x + 1$

4) In each of the above equations, what does the number before the "x" tell you?

5) Now, given what you have learned above, graph each of the following without making a table.

   a) $y = \frac{3}{2}x - 4$
   b) $y = -\frac{1}{3}x - 2$
   c) $y = -3x$

Two Forms

In general, there are two common forms for expressing linear equations. One is called *standard form*, where there are no fractions and the x's and y's are both on the left side, such as:
$$4x + 3y = 15$$
If we now solve this equation for y, then we get *slope-intercept form*, which for the above equation is: $y = -\frac{4}{3}x + 5$

— Cartesian Geometry – Part I —

Problem Set #3

On the previous problem we saw how if we have an equation solved for y, then the number in front of the x (which is called the *slope*) tells us how steep the line is, and the constant at the end tells us where the line crosses the y-axis (and this is called the *y-intercept*, and it is where the value for x is equal to zero).

1) Give the slope of all of the lines below.

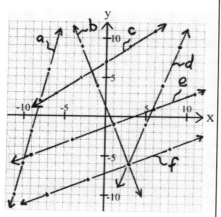

Questions about slope...
(It may be helpful to look at your answers to the previous problem.)

2) What does a negative or positive slope tell us about the direction of the line?

3) What is the slope of a line that is 45° (off horizontal)?

4) What can be said about the slope of a line that is less steep than 45°?

5) What can be said about the slope of a line that is steeper than 45°?

6) What can be said about the slopes of two lines that are parallel with each other?

7) What can be said about the slopes of two lines that are perpendicular to each other?

8) What is the slope of a line that is horizontal?

9) What is the slope of a line that is vertical?

10) Graph the following equations.
  a) $y = \frac{1}{3}x - 4$
  b) $y = \frac{3}{2}x - 1$
  c) $y = -\frac{3}{2}x + 1$
  d) $y = \frac{3}{4}x + 2$
  e) $y = \frac{1}{2}x$
  f) $y = -5x$
  g) $y = 2x - 5$
  h) $y = -3x + 2$
  i) $y + 3x = 2$
  j) $4y - x = -8$
  k) $y = 4$
  l) $3x + 2y = 2$

11) The last equation above is the same as what other equation given further above?

— Cartesian Geometry – Part I —

12) Consider the equation $y = x^2 - 4x$.
   a) Give three solutions to the equation.
   b) What are all the possible values that x can have? (In other words, is there a limit to how big or small x can be?)
   c) What are all the possible values that y can have? (In other words, is there a limit to how big or small y can be?)
   d) Graph the equation.
   e) Does graphing the equation give you any insights into the answers to part b and c?

13) Consider the equations
   $4y + 3x = 6$
   and $y - 2x = 7$
   a) Give three solutions to each equation.
   b) Find the common solution to the two equations by using algebra.
   c) Graph each equation (on the same graph).
   d) What is the common point on the graph?

Problem Set #4

1) Give the equation of each line both in slope-intercept form and in standard form.

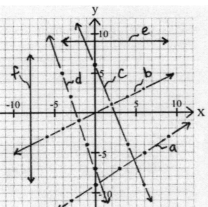

2) Give the equation of the line that…
   a) Has a slope of –3 and passes through (–2,7).
   b) Has a slope of ⅔ and a y-intercept of 5.
   c) Passes through (–5, –3) with a y-intercept of 2.
   d) Passes through the points (6,5) and (3,4).
   e) Passes through the points (6,5) and (–3,–7).
   f) Passes through the points (6,5) and (2,2).
   g) Passes through the points (6,5) and (–4,–3).

3) Give three other equations that have the same solutions as $y = \frac{1}{2}x + 3$

Three Methods
Here are the three common methods for finding the common solution to two linear equations:
- The *substitution method*
- The *graphing method* (as done at the end of the previous problem set.)
- The *linear combination method*. This method may be new to you, so here is an example:

Cartesian Geometry – Part I

Example: Use the *linear combination method* to find the common solution to these two equations:
2x + 3y = 4
3x − 4y = 23

Solution: We can choose to either have the x's cancel or the y's cancel. In this case, we will choose to cancel the x's. To do this, I multiply the top equation by 3, and the bottom by −2. So now the equations are:
  6x + 9y = 12
  −6x + 8y = −46
It is important to realize that these two equations are equivalent (i.e., they have the same solutions) as the original two equations.
  Here's the key: we simply add the two equations together, and the x's cancel. That's why we changed the equations to begin with!
  Now we have: 17y = −34
  This gives us y = −2 as a solution, and by substituting in for y, we get x = 5. The common solution to the two original equations is (5, −2).

4) Use each of the three methods to find the common solution to
  2y + 3x = −10
  6y − 5x = 26

Problem Set #5

1) Give the equation of each line both in slope-intercept form and in standard form.

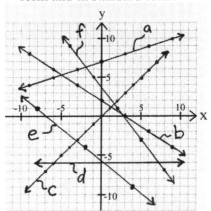

2) Graph each equation.
  a) y = ¾x + 2
  b) y = −2x + 7
  c) y + 2x = 7
  d) x + 2y = 7

  e) 3x − 5y = 10
  f) 3y + 2x = 5
  g) y = −x
  h) x = 4
  i) y = −x² + 6x − 9

3) Consider the equation 2x − 3y = 12.
  a) What is the slope of its graph?
  b) What is the y-intercept?
  c) What is the x-intercept?
  d) Where is x = −3?
  e) Give three solutions to the equation.
  f) For which point is the value of x and y the same?
  g) What solution does it have in common with x + 3y = 15

— Cartesian Geometry – Part I —

Temperature Conversions

4) The formula for converting from Celsius to Fahrenheit is:

$$F = \tfrac{9}{5} \cdot C + 32$$

a) What does the 32 indicate?
b) What does the $9/5$ indicate?

Use a full-size sheet of graph paper to graph the above equation. The vertical axis should be F, and the horizontal axis should be C. Both axes should have a range from −100 to 100.

Use this graph to estimate the answers to the following:
c) Convert 95°F to °C
d) Convert 10°C to °F
e) Convert 43°F to °C
f) Convert 43°C to °F

5) The formula for converting from Fahrenheit to Celsius is:

$$C = \tfrac{5}{9} \cdot (F - 32)$$

Multipying in gives us:

$$C = \tfrac{5}{9}F - 17\tfrac{7}{9}$$

a) What does the $17\tfrac{7}{9}$ indicate?
b) What does the $5/9$ indicate?

Use a full-size sheet of graph paper to graph the above equation. The vertical axis should be C, and the horizontal axis should be F. Both axes should have a range from −100 to 100.

Use this graph to estimate the answers to the following:
c) Convert 95°F to °C
d) Convert 10°C to °F
e) Convert 43°F to °C
f) Convert 43°C to °F

6) a) If the graphs from problems #4 and #5 are superimposed upon each other, where would the lines meet?
b) What is the significance of this meeting point?

7) Give the equation of the line that…
a) Has a slope of ⅔ and a y-intercept of (0,6).
b) Has a slope of −5 and passes through the point (2,−7).
c) Passes through the points (3,−2) and (−6,−5).
d) Passes through the points (10,6) and (5,4).
e) Passes through the points (10,6) and (4,2).
f) Passes through the point (3,−2) and runs parallel to the line $y = -2x + 9$
g) Passes through the point (−1,4) and is perpendicular to the line $y = ¾x - 5$

8) Use the linear combination method to find the common solution to:
$$5x - 6y = 31$$
$$3x + 4y = -8$$

9) Use each of the three methods to find the common solution to
$$x + 2y = 2$$
$$4x - 3y = 30$$

— Cartesian Geometry – Part I —
Problem Set #6

1) Give the equation of each line.

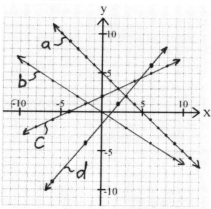

2) Graph each equation.
   a) $y = -\frac{1}{2}x$
   b) $y = \frac{3}{4}x - 5$
   c) $6y + 5x = 18$
   d) $3y + 7x = -18$

3) Use both the graphing method and the linear combination method to find the common solution to the equations given in #2c and #2d above.

4) Use both the graphing method and the substitution method to find the common solution to the equations given in #2b and #2c above.

5) Give the equation of the line that…
   a) Has a slope of $\frac{2}{3}$ and passes through the point (−6,1).
   b) Passes through the points (2,−5) and (6,−7).
   c) Passes through the points (1,7) and (−3,5).
   d) Passes through the point (10,4) and runs parallel to $y = -\frac{2}{5}x + 20$.
   e) Passes through the point (10,4) and is perpendicular to $y = -\frac{2}{5}x + 20$.

6) Jason currently has $3,400 of debt from an interest free loan, and he has a total of $400 in savings. He decides that starting today he will pay $100 per month toward his debt and that he will also save an additional $150 per month by putting it under his mattress (therefore no interest).
   a) Give an equation that expresses the balance of his debt over time.
   b) Give an equation that expresses his total savings over time.
   c) Graph the two above equations on the same graph.
   d) How much savings will he have after 2 years?
   e) When will he have $1600 in savings?
   f) When will his debt finally be zero?
   g) Where do the two graphs meet? What is the significance of that point?

About the Authors

Andrew Starzynski grew up steeped in Waldorf education as the son of two Waldorf teachers, and as a student at the Chicago Waldorf School, which he attended through high school. Andrew went on to study at Beloit College, graduating in 2001 with a double major in mathematics and philosophy, and a minor in computer science. He taught part-time at the Chicago Waldorf School before spending two years in the computer industry. In 2004, his love of math led him to graduate school where he acquired his master's degree in applied mathematics. His enthusiasm for Waldorf education prompted his return to the Chicago Waldorf School where he taught for three years. In 2008, Andrew completed his Waldorf teacher training under the tutelage of Jamie York. He then moved to Hawaii where he taught at the Honolulu Waldorf School until 2013. Currently he teaches at the Waldorf Academy in downtown Toronto where he resides with his wife Erika and young son Eoin.

Jamie York was born in Maine, went to public school in Connecticut, received two computer science degrees (from Rensselaer Polytechnic Institute in Troy, NY, and the University of Denver), and began teaching math in 1985 at a boarding school in New Hampshire. In 1994, after spending two years in Nepal serving as a Peace Corps volunteer, Jamie's search for meaningful education led him to Shining Mountain Waldorf School (in Boulder, Colorado), where he is still teaching middle school and high school mathematics. Since then, he has focused largely on envisioning and creating a comprehensive and meaningful mathematics curriculum that spans grades one through twelve. Jamie consults at a variety of schools, teaches math workshops, and serves on the faculty at the Center for Anthroposophy (in Wilton, NH) training Waldorf high school math teachers.

JAMIE YORK
PRESS

*Meaningful Math Books for Waldorf,
Public, Private and Home Schools*
www.JamieYorkPress.com